Frederick P. (Frederick Putnam) Spalding

Hydraulic Cement

It's Properties, Testing and Use

Frederick P. (Frederick Putnam) Spalding

Hydraulic Cement
It's Properties, Testing and Use

ISBN/EAN: 9783744649834

Printed in Europe, USA, Canada, Australia, Japan

Cover: Foto ©berggeist007 / pixelio.de

More available books at **www.hansebooks.com**

HYDRAULIC CEMENT.

ITS PROPERTIES, TESTING,

AND USE.

BY

FREDERICK P. SPALDING,

*Assistant Professor of Civil Engineering at Cornell University;
Member of the American Society of Civil Engineers.*

FIRST EDITION.

FIRST THOUSAND.

NEW YORK:
JOHN WILEY & SONS.
London: CHAPMAN & HALL. Limited
1898.

PREFACE.

THE following pages contain the results of a careful study of the nature and properties of hydraulic cement, and the various methods which have been proposed, or are in use, for testing cement.

The subject is not so simple as it might seem to the casual observer, but abounds in contradictions in the results of experiments and conflicts of opinion between authorities, which at times are quite bewildering.

The views of the author, as derived from his own observation of the behavior of cement in use or in the laboratory, have been stated without reserve, and free use has been made of the results of available European investigations. The recommendations of the recent commissions appointed in Europe for the study of the methods of testing materials are fully given in so far as they relate to cement.

The various tests applied for determining the quality of cement are discussed, and an effort is made to point out the limitations within which they may be accepted as reliable indications of value.

A chapter is given upon the use of cement in mortar and concrete, and a number of sample specifications are appended for the purpose of showing the present practice of leading American engineers.

F. P. S.

ITHACA, N. Y., March, 1897.

iii

CONTENTS.

CHAPTER VII.

TESTS FOR SOUNDNESS.

CHAPTER VIII.

SPECIAL TESTS.

CHAPTER IX.

CEMENT-MORTAR AND CONCRETE.

APPENDIX.

SPECIFICATIONS FOR THE RECEPTION OF CEMENT.

HYDRAULIC CEMENT.

CHAPTER I.

HYDRAULIC LIME.

ART. 1. DEFINITION.

LIME is the name commonly applied to the product obtained by the calcination of limestone. The limestones employed differ greatly in composition, and the properties of the limes obtained from them vary with the nature and proportions of the substances combined in them.

When the limestone is composed of nearly pure carbonate of lime, the clinker resulting from the burning, known as *quicklime*, possesses the property of breaking up, or *slaking*, upon being treated with a sufficient quantity of water. The slaking of lime is due to its rapid hydration when in contact with water, and the process is accompanied by a considerable increase in the volume of the mass of lime and by a rise in temperature. If the quantity of water be only sufficient to cause the hydration of the lime, the quicklime is reduced to a dry powder; while if the water be in excess it becomes a paste.

The slaked lime thus formed possesses the further property, when mixed to a paste with water and allowed to stand in the air, of hardening and firmly adhering to any surface with which it may be in contact. This hardening of common limes will take place only when exposed to the air and allowed to become dry.

When lime is nearly pure and its activity is very great it is known as *fat lime*.

If the lime have mixed or in combination with it considerable percentages of impurities of an inert character, which act as an adulteration to lessen the activity of the lime, causing a partial loss of the property of slaking and also diminishing its power of hardening, it is known as *meagre lime*.

When the impurities in the lime are composed mainly of silica and alumina, they may, while lessening or destroying its property of slaking, impart to it the power of hardening under water and of *setting* without reference to the presence of air.

When the proportions of the hydraulic ingredients are such that the material possesses the property of hardening in water, without having entirely lost that of slaking, the material is known as *hydraulic lime*.

If the acquirement of hydraulic properties has been accompanied by an entire loss of the property of slaking, the product is *hydraulic cement*.

Hydraulic limes and cements may be made either by burning limestones containing the proper proportions of hydraulic ingredients, in which case they are known as *natural* limes or cements, or by the admixture of material containing such ingredients to the limestone before burning, or to the lime afterward, in which case they are known as *artificial* limes or cements.

The materials used in the manufacture of lime and cement vary widely in different localities, and the resulting products differ greatly in their properties, being affected both by the composition of the raw materials and by the manipulation given them during the process of manufacture. Because of this variation in the character of the material, it is extremely difficult to formulate any general laws governing its properties, or to devise any system of testing which shall give an accurate determination of value.

Hydraulic limes are used quite extensively in Europe, but are not made to any extent in this country. The American cement industry is, however, a very important one. Natural cements are manufactured in immense quantities, and the production of Portland cement is rapidly growing to large proportions, although considerable of high-grade foreign cement is still imported.

ART. 2. CHEMICAL INGREDIENTS.

The most important ingredients of hydraulic limestones, in addition to the carbonate of lime, are usually alumina, silica, oxide of iron, and magnesia. They commonly contain also small quantities of sulphuric acid, phosphoric acid, oxide of manganese, potash, soda, bituminous or carbonaceous matter, organic substances and water. The lime after burning may also contain small particles of cinders from the combustibles employed.

The volatile substances are without effect upon the lime because they disappear in burning. A small percentage of carbonic acid and water which escapes being driven off in the burning or is afterward absorbed from the air will appear in the lime. If this quantity of carbonic acid be large, it indicates either that the burning has been incomplete or that the lime has become carbonated by subsequent exposure. The energy of the lime is thus diminished, the portion of lime in combination with the carbonic acid being rendered inert.

Silica is undoubtedly the most important element in rendering lime hydraulic and is always present. When in the form of silicious sand not attacked by acids it is unaffected by the burning and remains as inert material in the product. It is only that portion of the silica which is present in a condition to be reduced in the burning and combine with the lime which is of value in imparting hydraulic properties to the lime.

Alumina is an important element in hydraulic lime, although not, as seems to be the case with silica, an essential to its hydraulicity. Vicat found that alumina without silica would not render lime hydraulic. When, however, it is combined with silica it becomes one of the active elements in the setting and hardening of hydraulic limes and cements, provided it be not present in too large proportions. When in excess of about one part alumina to two parts silica, it is claimed that the surplus remains inert in the lime and detracts from its energy. This, however, very rarely happens in practice.

Oxide of Iron is commonly thought to be without influence upon the hydraulicity of lime, although there is some question concerning it. Like alumina, it confers no hydraulic properties when alone, and it is probably always inert.

Magnesia seems to act quite differently as the conditions under which it is present vary, and its real action in most cases is in considerable doubt. Vicat made hydraulic limes by burning carbonate of magnesia with fat lime, thus showing that the magnesia alone might impart hydraulic properties to the lime. It has also been shown by Mr. H. St. Claire Deville that the oxide of magnesium by itself is sometimes an hydraulic material and sets under water. Other experiments have seemed to indicate that magnesia might act like alumina, and replace that element in forming an hydraulic material with silica and lime. These results may possibly all be due to the hydraulic properties of the magnesium oxide and independent of any action of the magnesia upon the lime.

The silicates and aluminates of magnesia also have the property of hardening under water like the similar salts of lime, and in some cements a portion of the lime which would otherwise be required seems to be replaced by magnesia. The action in these instances is not definitely known, and there is a difference of opinion amongst authorities concerning it, some thinking that the magnesia acts like lime, others that it is inert and does not contribute to the energy of the material. In view of the activity of the magnesian salts when alone, it seems reasonable to suppose that they have a similar effect in the magnesian cements.

The activity of magnesia in hydraulic material depends largely upon the temperature at which it is burned, and in the experiments which have demonstrated its hydraulic properties it has been found that too high temperature destroys these properties. This possibly accounts for many of the contradictory results obtained by different investigators with these salts. This also is true in a less degree with the lime salts, which may sometimes be rendered inert by too high temperature.

Sulphuric Acid occurs in some limestones as sulphate of lime, and sulphur also sometimes occurs as a sulphide, usually of iron. During the burning the sulphide may become transformed into sulphate, and either or both forms may result in the final product. Experience seems to indicate that the sulphate is, in general, a deleterious substance, likely to affect the durability of mortar made from lime in which it occurs. The sulphide, however, is supposed to increase the hydraulic properties of the material, although its action is doubtful. In Europe it is common to limit the percentage of sulphur or sulphuric acid which may be present in the lime.

Phosphoric Acid occurs in very small quantities in material of this character, and is thought to have no action other than that of combining with a small quantity of lime, which is thus rendered inert.

Oxide of Manganese is comparatively unimportant, as the quantity present is always too small to have any considerable effect upon the properties of the lime. It has been supposed by some authorities to have strong hydraulic properties.

The *Alkalies* contained in the limestone act as a flux during the burning, causing the chemical reactions to take place more readily and completely. The small amount which ordinarily ocurs in lime and cement is unimportant in its effect, the alkali being gradually dissolved out of the mortar.

ART. 3. HYDRAULIC INDEX.

The hydraulic activity of a lime or cement, that is, its ability to harden under water, depends primarily upon the relative proportions of the hydraulic ingredients and of lime. Silica and alumina are considered to be the effective hydraulic ingredients, and it is common to designate the ratio of the sum of the weights of silica and alumina to that of lime in the material its *hydraulic index*.

The hydraulic index gives, therefore, within certain limits, a measure of the hydraulicity of the various classes of limes. It is to be remembered, however, that there are other factors to be considered in judging of the action of a lime than this simple proportion. The other ingredients may by their combinations withdraw portions of the active elements so as to modify the effective ratio between them, while the activity of the lime depends largely upon the state of combination in which the active elements exist. This is not shown by analysis, and may be greatly modified by the manipulation given the material during manufacture.

ART. 4. CLASSIFICATION OF LIMES.

Limes may be classified according to their hydraulic indices, or according to the rate of hardening under water.

Limes with hydraulic index less than 10/100 possess little if any hydraulic properties, and are known as *common limes*, which may be either *fat limes* or *meagre limes* according to the proportion of inert material contained by them.

When the hydraulic index is between 10/100 and 20/100 the lime is known as *feebly hydraulic* and may require from 12 to 20 days to harden under water. Hydraulic lime proper includes that of index varying from about 20/100 to 40/100. These may harden in 2 to 8 or 10 days, the more rapid ones being sometimes classed as *eminently hydraulic*.

When the hydraulic index is between about 40/100 and 60/100 the lime is of the class known as *limiting lime*. These limes have not the characteristics of hydraulic lime, but form the boundary between the limes proper and the cements, losing the property of slaking, and as a general thing not possessing a sufficient quantity of the hydraulic ingredients to make a safe natural cement. When burned at a low temperature it may give hydraulic lime of rather poor quality, and when properly treated and burned at a high temperature it makes good Portland cement. From material with an index of 65/100 to $1\frac{20}{100}$ are obtained most of the common *natural cements*. With hydraulic index of $1\frac{20}{100}$ to $3\frac{100}{00}$, the material yielded by burning is known as *meagre cement,* usually a weak material of little value as a cement.

The material of hydraulic index $3\frac{0.0}{100}$ may be puzzolana, which has not the properties of cement, but when mixed with fat lime renders the lime hydraulic.

The above divisions are all more or less arbitrary, and there are no sharp lines between the classes, which merge into and overlap each other. The character of the material is also affected by other factors than the hydraulic index, and classification by this method is by no means invariable, material of one class frequently behaving like that of another class. It is difficult to determine at exactly what point lime begins to be hydraulic, but when it requires a month or more to harden under water it is usually considered as being non-hydraulic.

In determining the rate of hardening of limes there are so many external circumstances which may affect the result that there is always chance for error, which causes classification by that method to be somewhat uncertain. Variation in temperature has always an important effect upon the rate of hardening.

ART. 5. COMMON LIME.

Common lime is such as does not possess hydraulic properties. It is divided into fat or rich lime and meagre lime, according to the quantity of impurities of an inert character it may contain. When made into a paste and left in air it slowly hardens. The process of hardening consists in the gradual formation of carbonate of lime through the absorption of carbonic acid from the air, accompanied by the crystallization of the mass of hydrated lime as it gradually dries out. In common lime the final hardening takes place very slowly, working inward from the surface, as it is dependent upon contact of the mortar with the air. When the lime is nearly pure the resulting carbonate is likely to be somewhat soluble, and consequently to be injured by exposure. Nearly all limes, however, contain small percentages of silica and alumina, and these ingredients, even when in quantities too small to render the lime hydraulic, impart a certain power to set, causing the hardening to take place with greater rapidity and without the same dependence upon contact with air. It also renders the material less soluble and more durable in exposed situations.

Limes containing but a small amount of impurities consist mainly of calcium oxide, which is very caustic and becomes hydrated very rapidly when brought into contact with water. This hydration, or slaking, produces a rise in temperature and an increase in volume, which vary in amount according to the purity of the lime, the volume being doubled or tripled for good fat lime.

The common method of slaking consists in covering the quicklime with water, using two or three times the volume of the lime. This method is known as *drowning*. The lime is usually spread out in a layer. perhaps 6 or 8 inches thick, in a mixing-box, the water poured over it and allowed to stand. Sufficient time must be allowed for all of the lumps to be reduced.

When the lime contains much foreign matter, the operation frequently requires several days.

Too great a quantity of water is to be avoided, the amount necessary being such as will reduce the lime after slaking to a thick pasty condition. All the water should be added at once, as the addition of water after the hydration is in progress causes a lowering of temperature and checks the slaking. For the same reason, the lime should be covered after adding water, and not stirred or disturbed until the slaking is completed. The covering is commonly effected by spreading a layer of sand over the lime; the sand being afterward used to mix with it in making the mortar.

A second method of slaking is sometimes employed having for its object the reduction of the slaked lime to powder and known as *slaking by immersion.*

Slaking by immersion is accomplished in two ways. By the first method, the lime is suspended in water in baskets for a brief period to permit the absorption of the necessary water, after which it is removed and covered until the slaking takes place and the lime falls to powder.

By the second method, sprinkling is substituted for immersion proper, the lime being placed in heaps and sprinkled with the necessary quantity of water, then covered with sand and allowed to stand.

The difficulty in using these methods is to get just the right quantity of water to make the slaking complete. Lime so slaked may be barreled and shipped in form of powder.

Spontaneous Slaking is also sometimes resorted to;

it consists in exposing the lime to the air until slaking is effected by absorption of moisture.

Lime slaked by immersion swells less than when slaked by drowning and requires less water to form into a paste.   Slaked lime, either as powder or paste, may be kept indefinitely if protected from the air.

Lime is commonly sold as quicklime, and should be in lumps and not air-slaked.   When it is old and has been exposed to the air, it is likely to have absorbed both moisture and carbonic acid, thus becoming less active, the portion combined with the carbonic acid being inert.   A simple test of the quality of quick-lime is to immerse a lump for a minute, then place in a dish and observe whether it swells, cracks, and dis-integrates with a rise of temperature.

Slaking some days in advance of use is desirable in order to insure the complete reduction of the lime, and it is quite common to slake lime much longer before it is to be used.

The swelling of lime has been found to be increased by slaking with steam instead of water, and slaking has been found by M. Candlot to be accelerated by the addition to the water of a small quantity (2% to 6%) of chloride of calcium or chloride of magnesium.

Common lime is ordinarily used in construction as a mortar, mixed with sand.   The quantity of lime in the mortar should be just sufficient to fill the voids in the sand, without leaving any part formed entirely of lime.   Mortar of rich lime shrinks on hardening, while masses composed entirely of lime on the interior are likely to remain soft, so that an excess of lime may be an element of weakness.   If too little lime be used, the mortar may be porous and weak. The proportions ordinarily required are between 1 part lime to 2 parts sand, and 1 part lime to 3 parts sand.

Mortar of common lime should not usually be employed in heavy masonry or in damp situations. Where the mass of masonry is large, the lime-mortar will become hardened only with great difficulty and after a long time. The penetration of the final induration due to the absorption of carbonic acid is very slow. The observations of M. Vicat showed that carbonization extended only a few millimeters the first year and afterward more slowly. The induration of the lime along the surfaces of contact with a harder material, when used in masonry, is usually more rapid than in the interior of the mass of lime itself, and hence the strength of adhesion to stone or brick is often greater than that of cohesion between the particles of mortar.

Meagre limes are similar in action to fat limes, but less energetic. They swell feebly in slaking and with slight change in temperature. The mortar hardens like that of fat lime, but cracks less and contracts less. Meagre limes proper, those containing so much inert matter as to materially reduce the energy, are not commonly employed in construction.

ART. 6. HYDRAULIC LIME.

Hydraulic lime is obtained by burning limestone containing silica and alumina in sufficient quantity to impart the ability to harden under water. The hydraulic elements are present in such quantities that

they combine with a portion of the lime, forming silicates and aluminates of lime, leaving the remainder as
free lime in an uncombined state.

When treated with water the free lime is slaked,
the action being much less energetic than that of fat
lime and varying in intensity with the quantity of
hydraulic ingredients.

The quantity of free lime in the material is dependent upon the degree of burning, as well as upon the
amount of lime contained by the stone.   If the stone
be underburned, the combination of the hydraulic elements with the lime is not complete, and more of the
lime remains in a free state.   For this reason a stone
of high hydraulic index may give a lime when underburned, but become unslakable when burned at a high
temperature, as in the case of the limiting limes.   The
best limes are usually those which can be burned at a
high temperature to complete the chemical combinations.

t is necessary that sufficient free lime be present
to cause the lime to slake properly, but it is desirable
also that the quantity of uncombined lime be as small
as possible, as the setting properties are due to the
silicates and aluminates, while the hydrated lime
remains inert during the initial hardening of the
mortar.

According to Professor Le Chatelier, limestone for
hydraulic lime should contain but little alumina, as the
aluminates are hydrated during the slaking of the lime
and become inert, while the silicates are not affected,
the heat of the slaking preventing their hydration.

The following is given as an average analysis of the best French hydraulic lime:

Silica.....................	22
Alumina...................	2.
Oxide of iron...........	1.
Lime......................	63.
Magnesia.................	1.5
Sulphuric acid...........	0.5
Water.	10.0
	100

It is important that the slaking be very thorough, as the presence of unhydrated free lime in the mortar while hardening is an element of danger to the work. Any lime becoming hydrated after the setting of the mortar may, by its swelling, cause distortion and perhaps disintegration of the mortar.

Hydraulic lime is used in the same manner as fat lime, being mixed with sand to a paste. When in the air hydraulic lime acts like common lime, dries, hardens, and slowly absorbs carbonic acid. It contracts and cracks when without sand, but much less than fat lime. In water, or in damp situations, the action of the two are altogether different. The hydraulic lime then hardens more or less rapidly. In running water a small amount of the lime is at first dissolved, but this is soon arrested as the hardening progresses.

Hydraulic lime is commonly slaked at the manufactory and shipped in form of powder. It may be kept, without injury, in this form by covering and protecting from the air.

ART. 7. MANUFACTURE OF HYDRAULIC LIME.

The manufacture of hydraulic lime as commonly carried on in Europe consists, after the quarrying of the rock, of burning, slaking, and bolting the material. As already stated, it is usually slaked at the works and sold in the form of powder.

The varieties of furnaces used in burning are quite numerous, but may be divided into those in which the stone is burned in contact with the fuel, and those in which the fire is outside the chamber in which the burning takes place.

Furnaces of the first class have been more generally employed, and are claimed (Candlot, Ciment et Chaux hydrauliques, p. 7) to be preferable from the point of view of the uniformity of product. Continuous furnaces are commonly used. The furnace is filled by placing alternate layers of combustible and limestone. When full it is lighted at the bottom, and as the mass settles, new layers of material are added at the top, while the burned lime is drawn out at the bottom, the furnace being kept in continuous operation. The rapidity of burning is controlled by dampers and the movable cover at the top.

In furnaces of the second class, the flame and gases of combustion are passed through the stone, which is not in direct contact with the fuel.

The regulation of a furnace to secure the proper degree of burning is a matter requiring skill and experience, and demanding the close attention of the attendant. The lime must be completely burned, but not overburned. The character of the limestone

determines the amount of burning necessary. A lime
of low hydraulic index may be burned at a much
higher temperature than one of high index. When
the limestone is irregular in character it will not burn
evenly, but the parts of high index will be vitrified
before that of low index is properly burned. The
burning is accelerated when the stone is moist, and
stone fresh from the quarry is preferred to that which
has been exposed until the hygrometric water has
been evaporated.

   The chemical phenomena of the burning of lime are
approximately as follows: The hygrometric water is
first driven off. The carbonate of lime is next decomposed. Then the clay is dehydrated and decomposed, and the combination of the silica and alumina
with the lime takes place. The temperatures required
to effect these changes depend upon the composition.
The decomposition of the carbonate of lime takes
place at a lower temperature in presence of silica, and
hence silicious limestones are burned at a comparatively
low temperature maintained for a considerable time,
while argillaceous limestones require a higher temperature maintained for a shorter period. According
to M. Bonnami, the carbonate of lime is commonly
decomposed at about 440° C., while the clay is dehydrated and decomposed at above 700° C., and the
silicates and aluminates of lime are then formed. If a
temperature intermediate between these points be
maintained for a considerable time, the result will be
the decomposition of the carbonate of lime, leaving
caustic lime mixed with insoluble clay.

   The *slaking* of hydraulic lime is commonly accomplished by sprinkling, as mentioned in Art. 5. The lime after coming from the furnace is spread in layers 4 to 8 inches deep in the slaking-chambers. It is then sprinkled so that all of the quicklime is well moistened; from 7% to 10% of water is commonly required. When the lime is wet it is thrown into large heaps and left for a sufficient time for the slaking to be completed. The time required varies with the hydraulic index and the degree of burning. Limes of high index may require 15 or 20 days. When the index and degree of burning increase together, the lime soon becomes unslakable.

In slaking, the object is to obtain the complete hydration of the uncombined lime without causing any change in the silicates. To accomplish this the slaking must take place at a high temperature. The heat of slaking volatilizes the surplus moisture and prevents the hydration of the silicates. If the temperature be too low, the lime may partially harden during slaking and the portion of silicate which is hydrated becomes inert. If the lime be imperfectly slaked, the free lime left in the material may cause injury to the mortar after hardening. Lime so affected will set more quickly than it would if sound, but afterward is likely to swell and crack.

Steam is sometimes used for slaking instead of water, and is said by M. Le Chatelier to act more rapidly upon the lime while producing no effect upon the silicates.

It is claimed that the aluminates are hydrated during the slaking of the lime, being readily acted upon by steam, and hence are undesirable in hydraulic lime.

After the lime has been reduced to powder by slaking it is forced through sieves which permit the passage of all pulverized particles, but hold those of appreciable size, including the underburned rock and the overburned parts which refuse to slake. The lime resulting from the first bolting is known as the *flour of lime*.

ART. 8. GRAPPIERS.

The residue left after the sifting of hydraulic lime is known as *grappiers*.

It differs very much in its composition in various instances, depending upon the limestone used and the manipulation in manufacture. It includes the underburned portions of the limestone and the overburned particles which will not slake. When the burning is thoroughly done and the limestone used regular in composition, the proportion of unburned particles is small. The larger part of the residue is then composed of hard material more rich than the other portions of the lime in silica and alumina, obtained from the clay which is disseminated through the limestone or formed by the combination of the cinders of combustion with the lime. This is what is properly meant by the term *grappiers*.

M. Bonnami found in his investigations that the larger part of the grappiers are from the surfaces of the limestone which are in contact with the combustible during the burning, and due to the silica and alumina of the cinders. These cinders are usually quite aluminous.

The grappiers are ground and sifted and either added to the lime or used separately as cement.

The addition of ground grappiers to hydraulic lime has the effect of raising the hydraulic index of the lime and increasing its activity, and offers a means of controlling to a certain extent the properties of the lime. In this case the mixture between the lime and the grappiers must be very intimate in order to obtain a homogeneous material. It is also very important that the lime in the grappiers be entirely slaked, to prevent the introduction of free lime into the product. To secure this the grappiers after grinding are exposed to the air for a considerable time before using, thus permitting any unslaked particles to become air-slaked.

The following analyses of lime and grappiers from the great works at Teil are given by Prof. Durand-Claye (Chimie appliquée à l'art de l'ingénieur).

ANALYSES OF LIME AND GRAPPIERS FROM TEIL.

	Slaked Lime.	Merchantable Lime.	Grappiers.	Rejected Material.
	Per Cent.	Per Cent.	Per Cent.	Per Cent.
Silica..................	23.05	23.95	31.85	43.90
Alumina and iron oxide.	2.75	3.10	4.25	8.20
Lime..................	65.75	63.35	55.60	45.25
Magnesia.............	1.50	1.15	1.20	0.85
Water, etc............	6.95	8.50	7.10	2.60

The first column gives the ordinary slaked lime. Merchantable lime has a portion of powdered grappiers added to augment its hydraulic properties. The third column gives an analysis of powdered grappiers, which is sold as a slow-setting cement. The rejected material is a calcareous sand which has puzzolanic properties. It is employed with cement in making water-pipe, brick, etc.

ART. 9. PUZZOLANA.

The term puzzolana is commonly applied to a class of materials which, when made into a mortar with fat lime or feebly hydraulic lime, impart to the lime hydraulic properties and cause the mortar to set under water.

Puzzolana (or pozzuolana) proper is a material of volcanic origin, deriving its name from Pozzuoli, a city of Italy near the foot of Mount Vesuvius, where its properties were first discovered. It was extensively used by the Romans in their hydraulic constructions, being pulverized and mixed with slaked lime and a small amount of sand for the formation of hydraulic mortar.

The puzzolana is a silicate of alumina in which the silica exists in a state easily attacked by caustic alkalies, and hence readily combines with the lime in the morta⁻

The class of puzzolanas also includes several other materials of somewhat similar character.

Trass is the name given to a volcanic material found in Germany and Holland, much resembling puzzolana and used in the same manner.

Arenes is a sand found in France and applied to the same purpose. It is quartzose in character and mixed with clay in considerable proportions; from 1/4 to 3/4 of the total volume. It may be made into a paste,

with water, which will harden on drying out, and is
sometimes used for common mortar without lime.

Psammites is a sandstone consisting of grains of
quartz, schist, feldspar, and mica, agglutinized with a
variable cement.   It is slaty in character and may be
worked into a paste with water.

Puzzolana may be made artificially by burning clay,
and natural ones may frequently be improved by burning, which has the effect of dehydrating the silicate of
alumina of which they are mainly composed and leaving it in condition to combine readily with the lime.

Berthier gives the following analyses of average
samples of puzzolanas:

	Trass.	Puzzolana.
Silica	0.570	0.445
Alumina	0.120	0.150
Lime	0.026	0.088
Magnesia	0.010	0.047
Iron oxide	0.050	0.123
Potash	0.070	0.014
Soda	0.010	0.040
Water, etc.	0.444	0.096

CHAPTER II.

CLASSIFICATION AND CONSTITUTION OF CEMENT.

ART. 10. CLASSIFICATION OF CEMENT.

HYDRAULIC cements may be classified according to the method of manufacture under five general headings: Portland cements, natural cements, slag-cements, mixed cements, and grappiers cements.

The term *Portland Cement* is commonly used to designate hydraulic cement formed by burning to the point of vitrifaction a mixture of limestone and clay in proper proportions and reducing the resulting mass to powder by grinding. The cement so classified is of lower hydraulic index than the other cements, and is consequently burned at a higher temperature. Portland cement is usually made artificially by a mixture of limestone and clay or of nearly pure limestone with stone of high index, and in all cases the material must be very uniformly incorporated into the mixture. The high temperature employed in burning and the necessity of reducing the raw material, whether natural rock or artificial mixture, to powder before burning, for the purpose of homogenizing it, may be considered the distinctive characteristics of this class.

The conference for the unification of methods for

23

testing materials at Munich * propose the following
additional definition of Portland cements: " They
contain a minimum of 1.7 parts of lime per unit of
hydraulic substances.   The addition of 2% by weight
of foreign matter may be tolerated in the manufacture
of Portland cement, with the view of augmenting certain important qualities, without the necessity of
changing the name.''

Natural Cements are those which are made by burning limestones less rich in lime than those giving
hydraulic limes or Portland cement.   These are
burned like the hydraulic limes without pulverization
of the raw material, and require a much lower temperature in burning than does Portland cement.

This class includes a number of sub-classes varying
widely in composition and value.   In Europe they
are commonly divided into quick-setting natural
cements, frequently called *Roman Cement*, and semi-
slow-setting cements, known sometimes as natural
Portland cement.   In the United States there is much
greater variety in the materials coming within this
classification and much confusion in their nomencla-
ture.   They are most commonly designated by a
name derived from the locality in which they are
obtained, and this seems the most feasible and satisfactory method.   Thus, Rosendale cements are those
from the region of the lower Hudson, Lehigh cements
are from Southeastern Pennsylvania, Louisville cements from the Ohio valley, Potomac and James River

---

* Mémoires de la Société des Ingénieurs Civils, 1891. vol. I.
p. 112.

cements from the corresponding valleys, while Utica, Akron, and Milwaukee are names indicating the location of manufactories of particular brands.

The term *American Cement* has sometimes been applied to include all natural cements made in the United States. This, however, often leads to confusion because of the fact that other than natural cements are now made quite extensively in this country, and the term *American Portland Cement* is also in common use

The term *Rosendale Cement* has frequently been given a general meaning and used as synonymous with natural and American to include all natural cements. It is, however, more properly restricted to the cements of the district in which it was first applied, and there seems to be no good reason for extending it to include other and totally different material.

Slag Cement or, as a more general term, *Puzzolana Cement*, is the product obtained by an intimate mixture of slaked lime with finely pulverized puzzolanic material, commonly blast-furnace slag. In this material the hydraulic ingredients are not burned with the lime, but are present in the cement in a mechanical mixture only.

Grappiers Cements are obtained by grinding the particles which are not pulverized in slaking hydraulic lime.

Mixed Cement is the name given in Europe to an extremely variable class of products obtained by mixing different grades of cement together, or by mixing cement with other material for the purpose of imparting desired properties.

ART. 11.   MANUFACTURE OF CEMENT.

The manufacture of hydraulic cement as commonly practised consists of four operations, viz., the preparation of the raw material, the burning, the grinding, and the bolting.

The methods of preparing the raw material differ according to the nature of the material and the method to be used in burning.   For natural cements it is usually only necessary to select the proper portions of the rock and break it into fragments of suitable size for introduction into the furnace.   The production of good cement requires the use of homogeneous material, and care must be used to prevent the introduction of variable rock into a single burning.

For Portland cement there are three general methods of preparing the material, in all of which it is essential to good results that the various ingredients be very carefully proportioned and that they be formed into a very uniform and homogeneous mixture in order to facilitate the chemical changes in all parts of the material during the burning.   The first, known as the wet method, consists in working the raw materials into an intimate mixture by reducing to a paste with water, then drying into bricks which may be stacked in the furnace for burning.   In the wet process proper a large excess of water is employed and afterward drawn off.   In the semi-wet process, now more commonly employed, only enough water is used to reduce the mixture to a plastic condition.

The second method, called the dry method, consists in grinding the materials together dry or with a very small quantity of water, ana making bricks of the powder by subjecting it to pressure in a brick-machine. The bricks in all cases require thorough drying before being placed in the furnace.

The third method is to grind the dry materials into powder and burn in a rotary furnace without forming them into bricks, or to mix to a plastic condition and dry in small lumps on the circumference of a drying-cylinder for burning in the rotary furnace.

The exact proportioning of the ingredients and the intimacy of their incorporation into the mixture have very important bearing upon the value of the cement.

The materials used in manufacturing cement differ greatly in different localities, and the method employed depends somewhat upon the character of the raw materials. For natural cements a limestone of high hydraulic index is usually employed, but differing much in composition, some having a high percentage of alumina, others of magnesia, and still others of both. For Portland cements the most common materials are a fat or slightly hydraulic limestone with clay or shale, made into bricks by the semi-wet process. Sometimes a hydraulic limestone of high index (such as is used for natural cements) is mixed with a fat limestone, commonly by a dry method. These materials are also sometimes used by the method of double calcination, that is, the fat limestone is first burned in the ordinary manner, the resulting quicklime is slaked and bolted, after which the slaked lime is mixed and ground with the argillaceous limestone, the object being to get a very perfect distribution of the lime through the mixture.

The method of double calcination is also sometimes employed with rock containing naturally about the right proportions of hydraulic ingredients for making Portland cement. The composition of the rock being usually somewhat irregular, it is lightly burned and reduced to powder, in order to secure greater uniformity, and then formed into bricks and burned in the usual manner.

The furnaces used for burning cement are of several kinds. The ordinary vertical lime-kiln, as mentioned in the previous chapter, is very commonly used, the kiln being used intermittently, and requiring usually 5 to 10 days to burn a charge.

The Hoffman continuous kiln is a series of chambers arranged in a circle, one chamber being fired at a time and the products of combustion passing through the chambers containing unburned material, the firing progressing from chamber to chamber continuously around the kiln.

The Dietzsch kiln is of the same character as the Hoffman, but has the fire outside the chamber containing the material to be burned.

The Ransome kiln consists of a rotary cylinder lined with fire-brick. The axis of the cylinder is inclined at an angle of 10 or 12 degrees with the horizontal, and the heat is applied as a gas- or oil-flame introduced through the axis of the cylinder at its lower end. The slurry is introduced in the form of powder at the upper end of the cylinder, and is slowly carried to the lower end by the corrugations of the inner surface of the cylinder, thus being gradually subjected to the heat, which reaches a maximum at the lower end, where the clinker falls out either into cooling cylinders or upon floors from which it is removed to storage-rooms.

The nature of the burning is indicated by the color of the clinker, which for Portland cement is dark green or black when well burned; underburned material being of a lighter color and weight, while the over-burned material may powder upon cooling, forming what is known as heavy powder. This heavy powder is inert and does not set as a cement, but ordinarily possesses puzzolanic properties and becomes active when mixed with lime.

The degree of burning required, as already indicated, varies with the nature of the material used, the heat required being greater as the hydraulic index of the material becomes less. The heavy powder mentioned above is formed at lower temperatures as the proportion of lime becomes less, and each grade of material has a certain range of temperature within which it should be burned, below which it will be underburned and above which it will be rendered inert.

In underburned cement the chemical changes are incomplete; a part of the lime may be left as caustic lime uncombined with the clay. This is shown by its light weight. In the burning, as the dehydration of the materials and the decomposition of the carbonate of lime is first effected, the limestone loses in weight without loss of volume, and thus suffers a loss in apparent density. As the subsequent combination of the lime with the clay occurs, a contraction in volume takes place and the density becomes greater.

After the burning is completed, the clinker is broken in a crusher to small fragments and then ground to powder. The cement is then bolted through sieves to separate the coarse particles, which are afterward returned to the grinders. The grinding is a very important matter, as only the extremely fine impalpable powder is of real value in the cement, and to secure good results the method of grinding must be such as to produce a large proportion of such material.

After the bolting of the cement it is usually carried to chambers and spread out for aeration, but in some cases is packed directly in barrels or bags for shipment.

The necessity for aeration depends upon the accuracy of the composition and the completeness of the chemical combinations. Its object is to eliminate the quicklime which may be present by allowing it to become air-slaked.

The method to be used in manufacturing cement must in each instance be modified to suit the material to be operated upon. The rock of a single quarry usually varies so much as to require different treatment in its various parts; or if mixtures are to be made, constant watchfulness is required in regulating the proportions in order to obtain a product of uniform quality.

ART. 12. PORTLAND CEMENT.

The term *Portland Cement* is usually limited to material containing a high percentage of lime and burned at a high temperature. It is usually low in alumina and magnesia. In order to make a good cement of this character it is necessary that the ingredients be very accurately proportioned and that the material be very homogeneous. This requires ordinarily the pulverization of the raw materials and their uniform incorporation into the mixture in a finely divided state.

The action of Portland cement seems to depend upon the formation, during the burning, of certain silicates and aluminates of lime which constitute the active elements of the cement, the other ingredients being considered in the light of impurities. The ideal cement would be that in which the proportion of lime is just sufficient to combine with all the silica and alumina in the formation of active material. If there be a surplus of clay beyond this point, it forms inert material. Any surplus of lime remains in the cement as free lime, and constitutes one of the chief dangers in the use of cement, as, although it may not prevent the proper action of the cement when used, it may cause the mortar to afterward swell and become cracked and distorted as the lime slakes.

As perfect homogeneity is not attainable in practice, it is always necessary that the clay be somewhat in excess in order that free lime be not formed. The amount of excess of clay necessary evidently depends upon the thoroughness of the process used in manufacture and the evenness which may be reached in the mixture of raw materials.

The hydraulic index of Portland cement varies from about 42/100 to 60/100. The value of the index is affected by the relative proportions of silica, alumina, and iron oxide contained by the cement as the equivalent weights of these oxides differ. The normal composition of Portland cement is usually within the following limits:

Silica...............	20	to	25 per cent.
Alumina...	5	"	9 "
Iron oxide...........	2	"	5 "
Lime	57	"	65 "
Magnesia........	0.5	"	2 "
Sulphuric acid........	0.25	"	1.50 "

Table I, taken from Candlot,* gives analyses of a number of representative European Portland cements, while Table II, collected from various sources, gives analyses of a few of the leading brands both American and foreign sold in this country.

A large number of analyses of European Portland cements are given by Professor Tetmajer † which show for the most part about the same range of variation as those already given.

Professor Le Chatelier has made a very careful study of the constitution of Portland cement by analyzing sections of clinker under the microscope, as well as by studying synthetically the various compounds of the principal ingredients. He concludes ‡ that the tricalcic silicate, SiO_4Ca_2, is the only one that is really hydraulic and is the active element in cement. In Portland cement he finds it to be the principal com-

---

* Ciment et chaux hydrauliques (Paris, 1891).
  Method en und Resultate der Prüfung der Hydraulischen Bindemittel (Zurich, 1893).
  ‡ Annales des Mines, September, 1893.

ponent, occurring in cubical crystals. It is formed by combination of silica and lime in presence of fusible compounds formed by the alumina and iron.

TABLE I.

COMPOSITION OF PORTLAND CEMENTS.

	Silica.	Alumina.	Oxide of Iron.	Lime.	Magnesia.	Sulphuric Acid.	Loss on Ignition.	Silicious Sand.	Not Determined.
French cements	22.20	6.72	2.28	67.31	0.95	0.26	0.40	....	....
	23.50	7.75	2.95	64.07	0.58	0.60	0.85	....	....
	21.70	7.48	2.57	65.54	0.90	0.77	1.20	....	....
	23.40	7.36	2.84	63.70	0.95	1.02	0.80	....	....
	24.50	7.09	2.81	62.40	0.85	0.70	1.25	....	0.40
	25.40	6.65	2.75	61.60	1.08	0.84	1.10	0.60	....
	21.80	6.56	2.64	57.42	0.72	0.34	0.40	0.12	....
	24.25	5.20	2.30	63.61	0.79	0.68	2.40	0.70	0.07
	22.30	8.04	3.71	58.68	2.20	2.23	2.55	0.25	0.04
	23 25	7.44	2.10	62.55	0.92	1.06	2.75	....	....
	23.00	8.33	3.87	60.90	1.10	1.40	1.49	....	....
	24.60	7.98	2.51	59.10	1.25	1.05	3.40	....	0.11
English cements	23.15	7.83	3.37	61.40	1.07	1.47	1.45	....	0.24
	23.30	7.65	3.10	62.20	1.04	1.06	1.60	....	0.05
	23.15	7.88	3.37	61.30	0.33	1.10	2.95	....	....
	23.70	7.80	3.40	59.36	0.55	1.25	4.10	....	....
	22.25	8.22	3.38	60.48	1.00	1.35	3.00	0.45	...
	21.95	7.99	3.91	59.08	1.04	1.52	4.35	0.35	....
	21.60	6.30	4.30	62.72	0.98	1.02	2.95	0.30	....
	21.35	7.15	3.75	62.16	0.95	1.06	3.20	0.25	0.13
	20.30	8.63	3.37	59.92	1.06	1.45	4.25	0.40	0.62
	23.30	8.13	2.67	60.48	0.60	1.20	3.90	....	....
	23.60	9.73	2 97	59.76	0.60	0.68	2.55	....	0.11
	24.05	8.69	3.31	59.69	0.90	1.47	1.85	...	0.25
	23 50	8.43	3.47	59.64	0.97	1.78	1.80	0.60	....
German cements	22.60	7.01	4.04	63.11	1.79	0.37	1.08	....	...
	21.75	8.16	3.64	63.39	2.30	0.51	0.25	....	....
	21.30	10.60	3.60	62.23	1.44	0.68	....	....	0.15
	24.85	6.07	2.43	64.40	1.26	0.51	....	....	0.48
	22.80	6.30	2.70	66.40	1.08	0.63	....	....	0.09
	23.70	5.25	2.70	67.18	1.00	1.40	....	....	....
	22.40	7.30	2.70	62.83	1.21	1.58	2.25	0.10	....
	22.80	7.46	2.84	63.28	1.24	0.98	1.55	0.20	....
	22.25	7.85	5.30	58.12	2.08	1.05	3.35	0.25	....
	20.80	8.66	3.64	62.52	1.68	0.89	1.85	0.10	....
Belgian cements	24.85	6.45	2 70	61.44	0.70	1.03	2.95	....	....
	24.50	8.51	2.84	60.03	0.88	1.54	1.20	0.60	....
	24.30	6.13	3.47	60.19	0.70	1.13	2.70	1.30	0.08
	23.80	6.39	2.51	62.32	0.72	1.17	2.94	....	0.14
	26.10	5.79	2.61	62.44	0.79	0.85	1.35	....	0.07
	24.30	5.33	2.67	64.12	0.72	0.74	1.95	....	0.17

TABLE II.

COMPOSITION OF PORTLAND CEMENTS.

	Silica.	Alumina.	Oxide of Iron.	Lime.	Magnesia.	Sulphuric Acid.	Alkalies.	Loss on Ignition.	Silicious Sand.
German .... {	21.14	6.30	2.50	66.04	1.11	0.73	0.67	1.13	0.28
	22.66	7.06	2.87	63.58	1.10	0.81	0.66	1.26	...
	24.90	11.	22	59.98	0.38	0.86	0.50	2.16	....
	20.32	7.86	3.60	61.66	1.85	1.31	1.20	1.07	0.80
English .... {	22.45	6.91	3.62	61.04	1.18	1.44	1.86	1.50	....
	23.65	10.56	3.34	60.00	0.97	1.12	....	0.50	....
	22.74	11.	44	57.68	0.51	0.60	0.63	5.40	0.53
	19.75	7.48	5.00	61.38	1.28	0.97	0.75	2.92	0.65
American ...... {	20.75	13.	50	62.25	0.25	0.25	2.25	0.47	....
	23.36	8.07	4.83	59.28	1.00	0.50	0.50	2.46	..
	22.45	13.	23	61.37	0.66	1.37	0.71	0.90	....
	20.80	8.60	3.70	62.51	1.09	1.29	...	2.05	0.10

" The bicalcic silicate, SiO_4Ca_2, possesses the singular property of spontaneously pulverizing in the furnace upon cooling. This silicate does not possess hydraulic properties and will not harden under water, but it is rapidly attacked by carbonic acid, forming carbonate of lime, and thus contributes something to the final hardening of mortar employed in air. The admixture of magnesia to form the double silicate of lime and magnesia, SiO_4MgCa, prevents the pulverization. This silicate is of no value for cement.

" At a very high heat the tricalcic silicate is decomposed into the bicalcic silicate and free lime, thus becoming inert."

" There are various aluminates of lime, all of which set rapidly in contact with water. The most important is the tricalcic aluminate, $Al_2O_6Ca_3$.

" With Portland cement a fusible silico-aluminate

of lime, $2SiO_2,Al_2O_3,3CaO$, is formed, identical with that which forms the essential element of blast-furnace slag, with a portion of iron replacing alumina. This compound is inert under the action of water and does not seem to be attacked by carbonic acid. Its useful function is to assist the combination of silica with the lime.

" This silico-aluminate, which is crystallized in Portland cement on account of slow cooling, is in a vitreous condition when the cooling is sufficiently brisk, as in the case of blast-furnace slag precipitated into cold water. It combines with hydrate of lime in setting, and gives rise to the hydrated silicates and aluminate of lime identical with those formed during the setting of Portland cement. It is these properties upon which are based the manufacture of slag-cement.

Prof. Le Chatelier gives two limits within which the quantity of lime in Portland cement should always be found. These are, that the proportion of lime should always be greater than that represented by the formula

(1) $$\frac{CaO + MgO}{SiO_2 - Al_2O_3 - Fe_2O_3} = 3,$$

and that it should never exceed that given by the formula

(2) $$\frac{CaO + MgO}{SiO_2 + Al_2O_3 - Fe_2O_3} = 3.$$

The symbols in these formulas represent the number of equivalents of the substances present, not the weights. One third the number of equivalents of sul-

phuric acid should be added to the denominator in
each case.

This is based upon the theory that the essential ingredients of the cement are the tricalcic silicate and
aluminate of lime and the silico-aluminate already mentioned.   Formula (1) represents the point at which
the amount of lime present would be just sufficient to
form the tricalcic silicate and the silico-aluminate, no
tricalcic aluminate being formed.   If less lime than
this be present, the bicalcic silicate would be formed.

Formula (2) represents the point at which the
amount of lime would be sufficient to form the tricalcic
silicate and aluminate to the exclusion of the silico-
aluminate.   If more lime than this be present, it will
remain in the form of free lime.

It is also stated by Prof. Le Chatelier that for Portland cement of good quality formula (1) usually gives
3.5 to 4, and formula (2) gives 2.5 to 2.7 as a result.

An examination of the analyses of a considerable
number of samples of good Portland cement shows
that in nearly all cases the requirements of the formulas
are met, and that most of them give results within the
limits specified above, but there are good cements for
which formula (1) gives considerably higher results
than 4.

Dr. Erdmenger considers * that the equations are
not borne out by experience, as they involve the
assumption that magnesia may be considered as lime.
It is also pointed out that the formation of the bicalcic
silicate depends upon the temperature of burning, and,

* Journal Society of Chemical Industry, XI. 1035.

according to Prof. Le Chatelier, the tricalcic silicate may be decomposed, and become inert at a sufficient temperature.

Dr. Erdmenger also states that the powdering upon cooling may in some instances be prevented by cooling suddenly, as by plunging into cold water, and that when so treated the material does not become inert.

When cement is burned in contact with the fuel, the composition is modified by the combination of the silica of the fuel with the lime. According to M. Bonnami a sort of grappiers is thus formed, as with hydraulic lime, particles being thus produced less basic than the rest of the cement, and of the character of puzzolana. This material is distributed through the cement in grinding and tends to slightly raise the hydraulic index. It is inert of itself, but may act like a puzzolana in combining with lime in the final hardening of the cement.

Portland cement when of low index and thoroughly burned usually sets slowly, but varies greatly in this respect, as the composition changes or the degree of burning is modified. It commonly gains its ultimate strength much more rapidly than natural cements.

Art. 13. Natural Cements.

The term *Natural Cement* is commonly employed to designate a large number of widely varying products formed by burning natural rock without pulverization or the admixture of other materials. These cements are usually of higher hydraulic index than the Port-

lands, and consequently more lightly burned.   The
index varies from about 60/100 to 150/100.

The quick-setting natural cements, or *Roman Cements* as they are called in Europe, are obtained by
burning, at a comparatively low temperature, argillaceous limestones of rather high index.   These
cements are usually characterized by a very rapid set,
and slowness in gaining strength subsequently.   The
feeble burning gives incomplete reactions, and the
formation of the silicates of lime is not so complete as
in the heavily burned Portland cements.   A considerable percentage of aluminate of lime is present, which
is the cause of the quick set, and there is usually a
strong proportion of sulphate of lime, which is regarded
as a necessary ingredient having the tendency to make
the set more slow, where it might otherwise be too
rapid for practical use.   Some unburned material is
also commonly present in such cements, remaining as
inert matter.   Material of this character becomes inert
when the temperature of burning is increased to the
point where the chemical reactions would become
complete, the heavy powder previously mentioned
being formed at a much lower temperature than in
cement containing a higher percentage of lime.

Table III gives results of analyses of a number of
the leading European Roman cements collected from
various sources, and showing the ordinary range of
variation in composition for good material.

The semi-slow-setting natural cements of Europe
are often known as *Natural Portland Cements*.   These
are often of a composition quite similar to Portland
cement, but usually have a higher hydraulic index and .

TABLE III.

COMPOSITION OF EUROPEAN ROMAN CEMENTS.

	Silicious Sand.	Silica.	Alumina.	Iron Oxide.	Lime.	Magnesia.	Sulphuric Acid.	Loss on Ignition.	Not Determined.
1	....	22.60	8.90	5.30	52.69	1.15	3.25	6.11	....
2	6 00	24.80	7.00	4.80	44.12	2.08	3.60	7.50	0.10
3	....	21.70	8.29	3.71	52.68	3.52	3.56	6.20	0.34
4	....	23.60	7.99	4.31	57.40	1.50	2.10	2.75	0.35
5	....	21.80	10.03	3.77	55.00	2.80	2.74	3.75	0.11
6	2.00	26.80	10.39	4.61	46.10	1.72	1.74	6.40	0.24
7	10.70	30 80	7.82	5.13	33.04	0.93	2.90	8.20	0.48
8	2.40	25.45	9 25	3.85	47.95	1.45	0.70	8 95	....
9	....	29.55	8.35	4.10	47.50	3.85	1.35	5.30	....
10	....	21.00	8.40	5.10	52.05	1.00	2 50	9.95	....
11	....	23.40	12.90	3.30	47.70	1.05	3.30	8.35	....
12	0.85	29.05	7.95	3 75	46.05	2.80	1.10	8.45	....
13	....	25.85	9.10	4.10	51.60	0.85	1.50	7.00	....
14	4.35	27.35	7.73	3.85	50.25	1.05	0.55	4.85	....
15	....	25.85	10.00	4.85	54.20	1.65	1.00	2.45	....
16	....	29.10	12.50	4.65	48.60	1.70	1.90	1.55	....
17	3.40	24 65	11.35	5.25	50.45	1.15	1.25	2.50	....
18	0.50	20.00	8.40	5.70	52.05	0.95	2.80	9.60	....
19	2.60	27.10	4.10	3.75	48.70	0.65	0.95	12.15	....
20	...	28.06	6.65	3.30	47.79	1.08	1.66	10.44	...
21	....	27.54	9.25	3.80	54.58	0.50	0.64	3.69	....
22	....	20.54	8.72	3.23	51.85	2.31	3.93	9.28	....
23	....	21.04	12.72	4.04	51.38	1.20	5.51	3.65	....
24	....	21.29	9.36	3.51	51.74	4.24	6.01	3.22	....
25	....	23 35	9.69	2.96	54.71	1.08	1.95	4.31	....
26	....	20.60	9.92	2.56	52.42	3.91	6.06	3.30	....
27	....	28.36	12.12	1.57	47.28	2.24	2.00	4.52	....
28	....	25.64	8.76	2.15	44.87	1.93	5.11	10.16	....
29	....	25.70	9.26	3.38	50.86	1.54	1.51	6.67	....
30	....	29 74	11.92	3.68	42.93	2.08	2.34	6.20	....
31	....	22.14	8.74	3.69	58.41	2.02	2.90	2.12	....
32	....	23.35	8 20	3.74	57.94	1.63	2.98	2.82	....

are given a somewhat lighter burning. They are, however, more heavily burned than the Roman cements. Limestones in nature are not so homogeneous as the artificial mixtures used in making Portland cement, and the proportion of lime cannot be so great as in the more homogeneous mixtures without danger of producing an objectionable quantity of free lime in the cement. The use of material of this character, therefore, requires much care in order to produce good results. As the hydraulic index becomes greater the homogeneity becomes less important, as free lime becomes less likely to occur and less dangerous, and irregularities only have the effect of increasing the quantity of inert matter, which causes mortar made from the cement to gain strength much more slowly than with Portland cement of low index. It is to be observed that the material spoken of as inert, and which delays the gain in strength in the early period of hardening, may not be altogether inert, and may contribute to the final strength of the cement, as it is of a puzzolanic character and perhaps ultimately combines with the hydrated lime in the mortar.

These cements occupy an intermediate position between the artificial Portland cements and the Roman cements, and may approach either in composition. In fact, the same raw material may frequently produce either—if burned lightly giving the quick-setting Roman cement, or burned more heavily a slow-setting natural Portland. Heavy burning increases the amount of silica combined with lime at the expense of the aluminates, thus relaxing the rapidity of set and increasing the early strength of the mortar.

The *Magnesian Natural Cements* are those in which a portion of the lime of the Roman cement is replaced by magnesia. Very little is known as to the action of the magnesia in these cements. It seems probable that the magnesia replaces lime or combines with it in the formation of double silicates and aluminates, and that it bears some part in the setting and hardening of the mortar. That certain magnesian salts possess hydraulic properties is well known, their action according to M. Fremy being probably much slower than the corresponding lime-salts.

The action of cements of this class is somewhat similar to that of Roman cements: they gain strength very slowly, but may be either quick or slow setting. The composition of the magnesian cements varies from that of the Roman cements to one in which the proportion of magnesia is as large as that of lime. As the proportion of magnesia to lime increases, the hydraulic index, considering magnesia as lime, frequently decreases and becomes less than would be admissible in Roman cement.

Magnesian cements are but little used in Europe, but in the United States they form the largest part of the natural cements in use, and many of them have been found by experience to be very useful and reliable materials. The Rosendale cements are of this character. The rock from which these cements are made differs greatly in character in the same locality, and in the different strata of the same quarry. In some of the works the nature of the product is regulated by mixing in proper proportions the clinker obtained by burning the rock from different strata. Each portion of rock must be burned in such degree as is suited to its composition, and hence as the material is not pulverized before burning it must be burned separately and mixed afterward. To produce uniformly good cement, therefore, requires close and careful attention; and for this reason there is often considerable difference in the quality of cement made by works in the same locality and from very similar material.

Cement of high index, unlike Portland cement, is usually materially changed by age.   When these cements are kept exposed to the air for a considerable length of time, they gradually become slower-setting and perhaps eventually lose the power of setting altogether, sometimes becoming puzzolana, which again becomes active cement by reburning.

Art. 14.   Slag-cements.

Slag-cement is formed by the admixture of slaked lime with ground blast-furnace slag.   The slag has approximately the composition of an hydraulic cement, but lacks a proper proportion of lime to render it active as a cement.   These cements are sometimes called puzzolana cements, the slag used being of the same nature as the puzzolana commonly employed in making lime hydraulic.

The method employed in forming slag-cement is to cool the slag suddenly by plunging it into a current of water as it emerges from the furnace.   This makes the slag granular, and causes it to retain the heat of crystallization, thus rendering its elements more ready to enter into combination in presence of water.

Experience in Europe shows that the slag must be basic in order to be of use in making cement.   Professor Tetmajer* arrives at the conclusion that slags in which the ratio $\dfrac{\text{lime}}{\text{silica}}$ is unity are not suitable, and that above this proportion the value of the product increases with this ratio. He also finds that the best results are obtained from slags giving a ratio

$$\frac{Al_2O_3}{SiO_2} = 45/100 \text{ to } 50/100.$$

M. Prost † states that a considerable amount of sulphur may be unobjectionable in slag-cements, and mentions a case where good results had been obtained with sulphurous slag, the only effect being discoloration attributed to sulphide of iron. He also concludes that a slag is most advantageous for this purpose which is most rich in lime and alumina.

It is very important in slag-cements that the slag be ground very fine, and be very intimately mixed with the lime. The lime is slaked and bolted, and then ground mechanically with the slag powder so as to insure thorough incorporation into the mixture.

In consequence of the necessity of attaining extreme pulverization of the slag, it is necessary to first dry it. The water which serves to make it granular remains to some extent between the grains and makes bad lumps at time of grinding. It has been attempted to substitute quicklime for slaked lime and use this water for slaking, but unsuccessfully, the slag combining to some extent with the lime and thus weakening the

---

* Annales de les Construction, Juillet, 1886.
† Annales des Mines, 1889, vol. II. p. 158.

cement, while particles of quicklime being left in the cement cause swelling of the mortar after setting. The drying is done in a furnace at a dull red heat.

The powdered slag is bolted through a fine sieve—about 10,000 meshes per square inch—before mixing with the lime.

The lime may advantageously be. kept for some time after slaking before being used, as this insures the complete reduction of the quicklime, but the slag seems to deteriorate when kept long after grinding. Fat lime is commonly employed for this purpose, but there seems to be an advantage in using meagre lime on account of the mortar being less likely to crack when used in the air. M. Prost found that there was no advantage to the strength of the cement in using hydraulic lime. Various additions of puzzolanic or other material are also sometimes resorted to for the purpose of preventing the cracking in air when fat lime is used.. This also increases the activity of the cement.

The composition of slag-cement usually differs from that of Portland in having a less quantity of lime, more silica and alumina, and more alumina in proportion to the silica.

Table IV gives the composition of a number of samples of the leading European slag-cements, taken from Candlot and Tetmajer.

Slag-cement is usually slow-setting, but the activity varies greatly with the circumstances of use. The rapidity of action is greater as the proportions of lime and alumina increase.

Slag-cement acts better under water than in the air.

TABLE IV.

COMPOSITION OF SLAG-CEMENTS.

	Silica.	Alumina.	Iron Oxide.	Lime.	Magnesia.	Sulphuric Acid.	Loss on Ignition.	Not Determined.
1	24.80	19.13	2.67	36.60	6.76	2.10	7.50	0.44
2	24.60	13.46	0.84	50.22	2.65	2.70.	5.40	0.13
3	24.90	13.46	2.83	50.40	1.20	1.10	6.45	....
4	24.30	13.85	1.15	49 50	2.16	1.86	6.90	0.28
5	27.45	14.65	1.75	46.20	1.86	0.72	7.00	0.37
6	25.20	15.23	0.77	50.00	1.35	0.72	6.50	0.23
7	20.40	18.59	0 41	50.07	0.50	0.08	8.30	....
8	18.30	18.07	0.34	53.16	0.64	0.18	8.07	....
9	22.35	12.83	0.64	55.61	2.17	0.27	4.01	....
10	27.35	9.13	1.50	50.28	5.72	0.40	2.59	....
11	20.35	14.05	0.33	50.26	2.68	2.39	6.99	....
12	18.69	9.20	2.14	46.36	4.92	1.25	12.19	....
13	19.87	14.84	0.80	48.54	2.44	1.00	8.41	....
14	18.11	15.54	0.92	54.72	0.54	0.37	8.64	....
15	20.94	14.85	1.03	48.18	3.58	1.69	7.22	....
16	19.24	17.15	1.07	54.21	0.81	0.39	6.39	....

It is essentially a hydraulic material, and it is especially important that it be kept damp during the early period of hardening, in order that the water necessary to its proper hardening may not evaporate.

M. Prost states that slag-cement is sensitive to the action of frost, and should not be used in freezing weather; while Mr. Redgrave declares that it resists frost better than Portland—showing a difference of experience in France and England.

Mr. Redgrave also says that it may be kept a long time without injury, and if kept free from moisture that it undergoes no change whatever; while M. Bonnami states that exposed to air in powder it rapidly loses its hydraulic properties, probably through carbonization.

ART. 15. MIXED CEMENTS.

The term *Mixed Cement* is sometimes used to include a considerable number of cements which are formed by a mixture of various products occurring at works where other cement is made. These mixtures may be made either for the purpose of cheapening the product or of imparting to it certain desired qualities. They consist of admixtures of different grades of cement, of the overburned or underburned portions of clinker, or of foreign material added to the cement.

Slag-cements and certain natural cements which, like some of the Rosendales, are made by mixing different grades of clinker are sometimes included under this head, but are not what is usually meant by the term.

Mixed cements differ so widely in character that no general discussion of their attributes is possible. Their values depend upon the care used in selecting, proportioning, and incorporating the ingredients, and each works has its own method of manufacture. In some cases, light-burned Roman cement is made slow-setting by the admixture of grappiers obtained in the slaking of hydraulic lime, with sometimes an addition of Portland cement. The overburned clinker from the manufacture of Portland cement is also sometimes utilized by being mixed with natural cement, a certain amount of Portland cement being added to bring up the initial strength and reduce the rapidity of set.

Cement of this kind is usually sold under the designation of Portland or natural cement, and not according to its real character. Some of them when carefully and regularly made give good results in practice.

ART. 16. GRAPPIERS CEMENT.

Grappiers cements are made by grinding to powder the grappiers left from the slaking and bolting of hydraulic lime. Very great care is necessary in eliminating all of the free lime from the grappiers by thorough slaking, the operation of slaking and bolting being repeated several times. The grappiers includes the underburned stone, and overburned material formed in contact with the fuel, as well as a certain amount of hard-burned material of too high hydraulic index to slake, and similar in composition and action to Portland cement. This latter is the effective portion of the cement, and it predominates in grappiers of good quality.

These cements are usually of low index and very slow-setting. They are liable to contain free lime unless carefully handled and usually require exposure to the air after grinding to permit them to become air-slaked.

At Teil the grappiers are passed through coarse grinders, which serve to remove all the soft parts. It is then bolted, allowed to air-slake for a month, then bolted again. Finally the parts resisting slaking are ground, steam being present to slake the particles of free lime, after which it is air-slaked before packing for shipment

ART. 17.   SAND-CEMENT.

Sand-cement is the name given to material formed
by grinding together Portlana cement ana sand to an
extremely fine powder and a very intimate mixture.
It is claimed that a very considerable amount of sand
may thus be mixed with the cement without materially
reducing its strength, and that the sand-cement so
made may still be mixed with the usual proportions
of ordinary sand and give good results in use.

It is said that the additional grinding given the
cement in pulverizing the sand reduces the cement to
impalpable powder, thus increasing its power of " taking sand."   Experiments also seem to indicate that
if sand be powdered separately, a certain amount may
be mixed with cement without serious injury to mortar made from the cement.

CHAPTER III.

THE SETTING AND HARDENING OF CEMENT.

ART. 18. THE SETTING OF CEMENT.

WHEN cement-powder is mixed with water to a plastic condition and allowed to stand, it gradually combines into a solid mass, taking the water into combination, and soon becomes firm and hard. This process of combination amongst the particles of the cement is known as the *setting* of the cement.

Cements of different character differ very widely in their rate and manner of setting. Some occupy but a few minutes in the operation, while others require several hours. Some begin setting immediately and take considerable time to complete the set, while others stand for a considerable time with no apparent action and then set very quickly.

The points where the set is said to begin and end are necessarily arbitrarily fixed, and are differently determined—usually by trying when the mortar will sustain a needle carrying a specified weight. The beginning of set is usually supposed to be when the stiffening of the mass first becomes perceptible, and the end of set is when the cohesion extends through the mass sufficiently to offer such resistance to any

49

change of form as to cause rupture before any perceptible deformation can take place.

It is sometimes stated that the chemical change involved in setting is an instantaneous occurrence at
about the time we call the beginning of set, and that
the gradual hardening then begins and is a continuous.
process until the maximum strength is reached. However this may be, with some cements a quite noticeable change suddenly shows itself at about this time in
the disappearance of water from the surface of the
mortar and the sudden stiffening of the mass.

ART. 19.   THE HARDENING OF CEMENT MORTAR.

After the completion of the setting of the cement
the mortar continues to increase in cohesive strength
over a considerable period of time, and this subsequent
development of strength is called the *hardening* of the
cement.

The process of hardening appears to be quite distinct from, and independent of, that of setting.   A
slow-setting cement is apt, after the first day or two,
to gain strength more rapidly than a quick-setting
one; but it does not necessarily do so.   The ultimate
strength of the cement is also quite independent of the
rate of setting.   A cement imperfectly burned may
set more quickly and gain less ultimate strength than
the same cement properly burned, but of two cements
of different composition the quicker-setting may be
the stronger.

There is as wide variation in the rate of hardening
of different cements as in the rate of setting: some

gain strength rapidly and attain their ultimate strength in a few days, while others harden more slowly at first and continue to gain in strength for several years. The rate of early hardening gives but little indication of the ultimate action of the cement, as the final strength of the mortar may be the same, however rapidly the strength is attained.

Portland cement usually hardens more rapidly and gains its maximum strength more quickly than natural cement, and also, as a rule, the Portland cement attains greater final strength when used in the same manner. Of two cements of the same class, however, it is not safe to infer that that which most rapidly gains strength will prove the stronger and more permanent material; in fact, where an abnormally high strength is shown in a few days the presumption as to final strength is against the cement giving such result, and in favor of one hardening at a more moderate rate.

The rate at which cement should harden for a given use depends, of course, upon the necessity of developing early strength in the work. For many purposes, such as most subaqueous construction, high early strength is quite desirable if not necessary; but for most engineering work a very rapid hardening does not seem necessary, and better results may often be obtained by the use of a material of more gradual action.

ART. 20. CHEMICAL THEORY.

Very little is definitely known concerning the chemical reactions which take place in the process of setting

and hardening of cement-mortars. Many theories
have been proposed to account for the phenomena by
different observers, based mainly upon the study of
the properties of various compounds of lime, silica,
and alumina formed synthetically. Chemical analysis
shows the proportions of the various elementary substances of which the cement is composed, but not
their state of combination; and the action of a cement
may be greatly modified by altering the condition in
which the ingredients exist, through changing the
manipulation in manufacture, without altering their
relative quantities.

M. Fremy considered Portland cement to be very
complex in composition, and ascribed the setting to
the action of lime upon certain puzzolanic compounds,
composed of double silicates of lime and alumina, the
calcination of the clay giving rise to a porous material
which absorbs the lime by capillary affinity.

M. Landrin concluded that a substance corresponding to the formula $3SiO_2, 5CaO$ is found in both Portland cements and puzzolana, and he considered this to
be the active element in the hardening of cement,
although he states that aluminate of lime contributes
to the setting and accelerates that action.

Prof. Le Chatelier, from his study of Portland
cements, explains the phenomena of setting by showing that certain salts, including the aluminate and
silicate of lime which form the active elements of
Portland cement, while soluble in an anhydrous state,
form insoluble salts when hydrated. When they come
into contact with water in mixing mortar the anhydrous salt enters into solution, then, becoming hy-

drated, the hydrate is precipitated from the saturated solution in a crystalline form. Those salts which are thus capable of being dissolved in an anhydrous state and then becoming hydrated arrive at stability in two ways—by decomposition and by combination.

The tricalcic silicate, which is the essential element of Portland cement, is decomposed in presence of water to a hydrated monocalcic silicate and a hydrate; thus

$$SiO_4Ca_2 + Aq = SiO_3, CaO, 2.5H_2O + 2CaO, H_2O.$$

The monocalcic silicate crystallizes in the form of needle-like crystals and the hydrate in hexagonal lamina visible to the eye. The tricalcic aluminate is hydrated by simple combination with the water.

$$Al_2O_6Ca_3 + Aq = Al_2O_3, 3CaO, 12H_2O.$$

The double silicate of alumina and iron, $2SiO_2$, $Al_2O_3, 3CaO$, is thought to be quite inert in Portland cement, and to merely serve the purpose of assisting the combination of the silica and lime by acting as a flux during burning. It seems, however, to be an active element in slag-cement, forming by combination with lime in presence of water the same compounds that are produced in the setting of Portland cement. The difference in its action is explained by the fact that in the slow cooling of Portland cement the salt exists in crystalline form, while through the sudden cooling of the slag it is made vitreous, and is then in condition to be attacked by the lime.

The first setting of Portland cement is attributed to

the hydrating of the aluminates and ferrites of lime, while the subsequent hardening is due to the slower progress of the hydration of the tricalcic silicate. The rapidity of set is therefore dependent upon the relative proportions of aluminates and silicates. When the burning is done at a low temperature, therefore, the aluminates, which are first formed, will cause a rapid set, while as the degree of burning becomes greater the aluminates give place to silicates which cause the setting to become slower and the subsequent gain in strength greater. The aluminates are thought to add but little to the final strength of the mortar, as they are not permanent compounds, but are acted upon by water and various salts with which they are likely to come in contact in the work.

Cements of low hydraulic index harden more rapidly and gain their full strength earlier than those of high index. They are more nearly of the composition which, according to the theory, should give the highest proportion of active ingredients, while those of high index have a surplus of silica and alumina, forming inert material. It is perhaps questionable whether in all cases this so-called inert material is in reality inert in the final hardening of the cement. Sometimes those cements which from this cause harden very slowly continue to gain in strength over a long period, and ultimately surpass those which gain strength more rapidly in the beginning; and it is quite possible that this overclayed portion, which is of puzzolanic character, may bear an important part in the final hardening.

Most slow-setting cements have a period during

which they lose strength after hardening for several months, probably due to the decomposition of salts formed by the parts of too low hydraulic index during the burning.   This loss of strength is usually temporary when the cement is of normal composition; but if it be overlimed, the loss of strength may continue, to the final destruction of the mortar.

The experiments of M. Candlot indicate that the presence of carbonic acid is essential to the hardening of hydraulic cement-mortar.   He found that if the mortar were placed in distilled water, frequently renewed, it became gradually decomposed, and finally lost all coherence; but the presence of carbonic acid, as is common in all natural waters, prevented this action and caused proper hardening to take place.

ART. 21.   INFLUENCE OF CALCIUM SULPHATE.

The action of sulphate of lime to slacken the rate of setting of Portland cement is well known.   In Germany it has been common to utilize it, for the purpose of regulating the rate of set, by adding powdered gypsum to the cement.

M. Candlot * has made a careful study of the influence of sulphate of lime upon the action of Portland cement.   He found that the increase in time required to set varied with the quantity of sulphate added; an addition to a quick cement of from 1 to 4 per cent being sufficient to change the time of set from a few minutes to several hours.   Cement which has been

* Ciment et Chaux Hydrauliques (Paris, 1891).

made slow-setting by the addition of calcium sulphate becomes again quick-setting with age, more or less rapidly, as it is or is not exposed to the air. In some cases the cement by exposure soon becomes quick-setting, and then by longer exposure again becomes slow. When cement treated with the sulphate of lime has regained its quick action by exposure, it may again be made slow by the addition of a small quantity of lime. Fresh cement with sulphate of lime added, and setting slowly in consequence, will set rapidly if the mortar be mixed with a solution of the carbonate of soda.

"Cement having sulphate of lime added set more rapidly when mixed with sea-water than with fresh water, and that which had been exposed and regained its former activity set more rapidly when mixed wet than when mixed stiff.

"The addition of a small quantity of calcium sulphate to Portland cement augments the strength. When the mortar is kept in sea-water and the proportion of sulphate exceeds 1 or 2 per cent, the mortar cracks and perhaps disintegrates. When the cement containing the sulphate was kept in sacks during several weeks it showed less strength during the early period of hardening."

M. Candlot concludes from his experiments that the explanation of the action of sulphate of lime lies in the fact that in the presence of water the sulphate combines with the aluminate of lime forming the compound, $Al_2O_3, 3CaO, 2.5(SO_3CaO)$, which crystallizes with a large quantity of water. The action depends upon the fact that the aluminate is insoluble

in lime-water, and, as most of the quick-setting
cements contain a certain quantity of free lime, when
the cement is gauged the lime at once enters into
solution and prevents the action of the aluminate until
the sulphate is combined with it. When the cement
becomes old, the free lime becomes carbonized, and
fails to prevent the immediate solution of the aluminate.

Aluminous cements burned at low temperatures
often contain considerable aluminate of lime, and these
may bear an addition of 5 to 10 per cent of sulphate
without loss of strength. The proportion of sulphate
must always be limited to what may be neutralized by
the aluminates.

ART. 22.   INFLUENCE OF CALCIUM CHLORIDE.

M. Candlot has also made a careful study of the
effect upon the setting and hardening of cement-mortar of chloride of calcium, either dissolved in the water
with which the cement is mixed, or that in which the
mortar is immersed. He found that Portland cement
gauged with water containing a few grammes per litre
of chloride of calcium sets more slowly than if gauged
with pure water; while if the solution of chloride be
concentrated, 100 to 400 grammes per litre, the setting is very rapid.

" The influence of calcium chloride in weak solution
upon the set of Portland cement may be attributed
to the salts which determine the set entering into solution more slowly in that solution than in pure water.
Hydrate of lime agitated with a large excess of water

is dissolved much less in the chloride solution than in pure water, while with the aluminate of lime this result is much more marked.

" If cement of Vassy, quick-setting, be gauged with a solution of chloride of calcium, 20 to 40 grammes per litre, the setting is about the same as with pure water. If the cement be diluted to a fluid with the same solution it will not set or harden. Portland cement treated in the same way hardens very slowly, but acquires a hardness comparable to that given by fresh water.

" Feeble solutions of chloride of calcium have no appreciable effect upon cements exempt from alumina, like certain grappiers cements composed almost exclusively of silicate of lime.

" From this the conclusions are drawn:

" 1. That in Portland cement the aluminate exists in feeble proportions; that it acts in an energetic manner upon the set, but very little upon the hardening, which is caused by the silicate of lime.

" 2. That in the Vassy cement the aluminate of lime is the essential element, and determines both the setting and the hardening; the rôle of the silicate being unimportant, especially during the early period of hardening.

" 3. That in the phenomena of setting the relative quantities of the elements present determine the action. The solution of chloride in presence of a large quantity of the aluminate of lime perhaps does not hinder the hydration and crystallization; but if, on the contrary, a small quantity of aluminate be mixed in an

excess of chloride solution, the action of that prepon-
derates, and the aluminate will not enter into solution.

' A weak solution of CaCl has the property of pro-
voking the rapid hydration of lime. A cement containing an excess of free lime, gauged with pure water,
swells and disintegrates under the slow expansive
action of the free lime. The same cement gauged
with a solution of CaCl, 30 to 60 grammes per litre,
does not swell, because the lime is slaked before the
set.

" As already stated, when Portland cement is mixed
with a solution of 100 to 400 grammes per litre CaCl
it sets very quickly. This set is accompanied by a
strong rise in temperature. This only occurs with a
fresh cement. With an old cement the setting remains slow, no rise in temperature is produced, and
the mortar swells and disintegrates.

" Mortar of cement gauged with a concentrated
solution of CaCl is disintegrated if placed in water
immediately after setting, but 15 or 20 hours afterward it may perhaps be submerged without loss of
strength.

" The action of a concentrated solution CaCl upon
Portland cement is due to the fact that aluminate of
lime is attacked very energetically by that solution.
While it is very slightly soluble in a feeble solution,
it is dissolved in large quantities in a concentrated
solution.

" When a fresh cement is agitated with a concentrated solution of CaCl it dissolves not only the aluminate, but the oxide of lime. The lighter the cement
is burned, the more it will dissolve. When an old

cement is agitated with the concentrated solution of CaCl the aluminate dissolves but very little.

ART. 23. EFFECT OF SAND.

Cement is ordinarily employed in mortar formed by mixing it with sand, and the action of the mortar is necessarily largely affected by the nature and quantity of sand used.

When the cement is finely ground and the sand of good quality, a mortar composed of equal parts of each, as a general thing, finally attains a strength as high as, or higher than, the neat cement. Cements of different characters, however, vary considerably in their power to " take sand " without loss of strength; some of the weaker ones may not be able to take more than half their weight of standard sand, while others can be mixed with considerably more than their own weight without loss of strength at the end of one year after mixing. All have a certain limit within which they may be made stronger by an admixture of good sand than they would be if mixed neat.

Cement mixed with sand always hardens more slowly than neat cement, and requires a much longer time to attain its maximum strength. As the proportion of sand to cement is increased both the rate of hardening and final strength are diminished. Within certain limits, however, the strength of mortar increases over a longer period of time as the proportion of sand becomes greater, and as the time of observation is extended the loss of strength due to larger proportions of sand becomes less. Thus a good Portland cement in a mortar containing 1 part sand to 1 of cement at the end of a year may be expected to be stronger than mortar of neat cement. At the end of three years the 1 to 1 mortar should be much stronger, while a 2 to 1 may be as strong as the neat mortar. At the end of four or five years the 2 to 1 mortar may be on even terms with the 1 to 1, while a 3 to 1 mortar may have steadily gained to perhaps three fourths the strength of the others, where it probably stops. Beyond the limit where the quantity of cement is sufficient to fill all the interstices in the sand the ultimate strength diminishes very rapidly as the proportion of sand increases.

Clean and sharp sand usually gives a higher strength in mortar than that containing an admixture of clay or earth, or that composed of rounded grains. Coarse sand also usually gives greater strength than that which is very fine. It is often difficult, however, to judge of the quality of sand without experimenting with it. In some cases a small amount of fine clay does not appear to injure the strength of the mortar, while a judicious mixture in the sand of grains of various sizes may be of benefit, through reducing the volume of interstices.

A mortar composed of sand and cement usually possesses greater ability to adhere to other surfaces when coarse sand is used than if the sand be fine.

ART. 24. WATER USED IN GAUGING.

The quantity of water used in mixing the mortar is one of the most important conditions; the less the

quantity, provided there be sufficient to thoroughly
dampen the mass of cement, the quicker will be the
set.   With some Portland cements, changing the
quantity of water used in mixing neat cement from 20
per cent to 25 per cent of the weight of cement
doubles or even triples the time required for the
mortar to set.   In other cases the effect is comparatively slight.

When the quantity of water used in mixing is sufficient to reduce the mortar to a soft condition the
hardening as well as the setting becomes more slow,
and the strength during the early period is less than
if a less quantity be used.   This difference disappears
to some extent with time, and the mortar mixed wet
may eventually gain nearly as much strength as though
mixed with less water.

When the quantity of water employed is not sufficient to reduce the mass to a plastic condition, the
mortar will not be so thoroughly compacted, and will
not reach the same strength as when made plastic,
unless pressure be applied to it.   But if just sufficient
water be used to thoroughly dampen the mortar, and
pressure be applied to expel the air and close the
voids, the early strength will be greater than when
more water is used.   This difference, like the former
one, disappears to a certain extent with time, but the
final strength is usually greater with the less quantity
of water.

According to Prof. Le Chatelier, the solidity of the
crystalline mass varies with the form, dimensions, and
mode of aggregation of the crystals.   In general, the
strength of a single crystal is greater than its adher-

ence to neighboring crystals. Long needle-like crystals give greatest strength, and crystals have this character more as the solution is more strongly supersaturated.

The nature of the water used in mixing may also affect the rate of setting. When sea-water is used the setting is usually slower than with fresh water, the chloride and sulphate of magnesia being the principal retarding elements. Cements with a high hydraulic index show a less difference between fresh and sea water than those of low index, and well-burned cements less than imperfectly burned ones. The experiments of M. Candlot indicate that this is due to the action of the salts mentioned above upon the aluminate of calcium, and that those cements containing the highest percentage of aluminate are affected the most by being mixed with sea-water.

Water containing sulphate of lime in solution retards setting (see Art. 21).

Mortar kept immersed in sea-water usually hardens more rapidly than that kept in fresh water. This difference is commonly much more noticeable with neat cement than with mortar containing considerable proportions of sand. The strength gained in sea-water, however, although gained much more quickly, is generally less in final amount than that in fresh water. There is, however, a very great difference between various cements in this particular.

Cements with a low hydraulic index show the greatest difference between sea and fresh water. Those containing small quantities of free lime give much greater early strength in sea than in fresh water, but are also sooner disintegrated by the action of sea-water.

ART. 25.   EFFECT OF ENVIRONMENT.

Cement-mortar kept under water ordinarily hardens more rapidly in the early period than that exposed to the air, but usually that kept in air ultimately reaches greater strength.   The highest strength is commonly produced by keeping the cement for a time in water, and later placing in dry air.   Nearly any cement-mortar will harden more rapidly and attain greater strength if kept moist during the operation of setting and the first period of hardening than if it be exposed at that time to dry air.   A sudden drying out about the time of completing the set usually causes a con. siderable loss of strength in mortar, and frequently the mortar so treated is filled with drying cracks. This result is usually more marked when the cement is mixed with a large quantity of water to a soft con- dition.

The nature of the water in which the mortar is allowed to harden is of more importance to its strength than that of the water used in gauging.   When the mortar is to be kept in air, the nature of the water used in mixing becomes more important, although probably the variations in ordinary natural water are rarely sufficient to produce any appreciable difference in the strength of the mortar.   Mortars gauged with sea-water harden best in air.

ART. 26.   EFFECT OF TEMPERATURE.

The temperature of the water used in mixing has an important bearing upon the time required for setting;

the higher the temperature, within certain limits, the more rapid the set. Many cements which require several hours to set when mixed with water at a temperature of 40° Fahr. will set in a few minutes if the temperature of the water be increased to 80° Fahr. Below a certain inferior limit, ordinarily from 30° to 40° Fahr., the mortar sets with extreme slowness or not at all; while at a certain upper limit, in some cements between 100° and 140° Fahr., a change is suddenly made from a very rapid to a very slow rate, which then gradually decreases as the temperature increases, until practically the mortar will not set.

The temperature of the cement, and that of the air in which the mortar is placed during setting, influence the rate of setting in about the same manner as that of the water. In case the air in which the mortar is placed be dry, the setting will usually be somewhat more rapid than if it be moist; and if it be too dry, the rapid evaporation of the water from the surface of the mortar may cause drying cracks in the mortar.

Quick-setting cements usually show a rise in temperature during setting, due to the rapidity of the action which takes place. It has been suggested that the time occupied by the setting would be better shown by observing the period of advanced temperature, than by noting the stiffening of the mortar, as is common. Most slow-setting cements, however, do not show sufficient change of temperature, if any takes place, to be appreciable; and the rise in temperature, where it does take place, may not always be the result of the process of setting.

If the air at the time of mixing mortar be sufficiently cold to freeze the mortar before it can set, it will not set while frozen; but most cements will do so after thawing out, and but few of them will be injured by such freezing in so far as their ultimate strength is concerned. Recent experiments * have, however, seemed to show that mortar may set while frozen if it remains in that condition for a sufficient length of time.

The temperature of the water with which cement-mortar is mixed has a quite appreciable effect both upon its rate of hardening and its ultimate strength, and the temperature of the air at the time of mixing has a similar effect. The lower the temperature at which the mixing is done, the slower the hardening and the greater the final strength. This difference is not sufficient to be important at ordinary air temperatures in so far as the use of mortar is concerned, but is quite appreciable in making comparative tests.

The temperature of the air or water in which the mortar is immersed during the time of hardening has a very appreciable effect upon the rate of hardening of many cements. This effect differs very radically for different material; with some the process is greatly accelerated by keeping them hot as compared with what would be the result in cold air or water; others are not appreciably affected, while still others seem to be retarded in their hardening by the application of heat. This variation is to be found among cements of the same class, and is seemingly independent of their value. Cements of low hydraulic index usually show the greatest gain in rate of hardening under the action of heat.

---

* Paper by Cecil B. Smith; Trans. Canadian Soc. C. E.

ART. 27. EFFECT OF AGE UPON CEMENT.

The effect upon a cement of retaining it a long time before using depends upon the nature of the cement and the method of keeping. When the cement is enclosed so as to prevent the access of air, as in barrels, it may usually be preserved for a considerable time without experiencing any alteration, provided it is kept dry.

When exposed to the air the cement commonly undergoes more or less alteration. Portland cements of good quality are usually but slightly affected, as they are composed for the most part of stable compounds. The change which occurs is limited usually to making the cement slower-setting. Where the cement is originally slow-setting this effect may be very slight, and the cement may perhaps be retained for a long period, two or three years at least, without appreciable change in its properties. A hard-burned cement, originally quick-setting, usually becomes slower-setting with age, but commonly without injury as to its ability to harden and its ultimate strength.

Light-burned cements, particularly Roman cements, are affected in much greater degree by age. These cements not only become slower-setting when exposed to the air, but commonly also they gradually lose the power of hardening and become finally inert—in many cases becoming puzzolanic material, the activity of which may be restored by the addition of slaked lime.

The changes which occur in cements kept in dry air are attributed to the action of carbonic acid upon the

free lime which they may contain, and perhaps also
upon the less stable compounds, as the aluminates of
lime, which contribute to the rapid set and are found
most plentifully in the light-burned material

ART. 28. EFFECT OF FINENESS.

The degree of fineness to which the cement is ground
is always very important in its effect upon the strength
of mortar made from the cement.   The valuable part
of the cement is practically only that portion which is
ground extremely fine—to an impalpable powder.
The coarse parts are not altogether inert, but are more
or less active, depending upon the size of the grains
of which they are composed.   As the clinker obtained
in burning cement is very dense and hard, it is ground
with difficulty, and the coarser particles are apt to be
of the best burned, and therefore most valuable, part
of the material.   Coarse grinding is likely therefore to
leave in a useless condition much of what should be
the most active portion of the cement.

The rate of setting is accelerated as the fineness to
which the cement is ground becomes greater.   In a
finely ground cement the amount of active material is
greater than in one coarsely ground, and the chemical
reactions which take place in setting are facilitated by
fine subdivision of the particles.   When the cement
is gauged with sea-water the rate of setting is less influenced by the fineness.

The hardening of cement mixed neat is not greatly
affected by fineness; that finely ground usually hardens more rapidly, but attains less final strength than

when more coarsely ground. The hardening of the coarsely ground cements is more gradual and regular, and the ultimate strength greater.

Cement, when used, is commonly mixed with sand, and the attainment of strength in sand-mortar rather than mortar of neat cement is therefore of importance. The finer ground the cement, the greater its resistance when mixed with sand, both in the earlier and later stages, and also the sooner will it reach its ultimate strength. The effect of fine grinding is much greater when the proportion of sand to cement is large, as the power of the cement to take sand without diminution of strength is thereby greatly increased. The coarser particles of the cement may be considered as practically inert material, which acts rather as sand than as cement in the mortar, and the power of the cement to harden and develop strength in sand-mortar is thus dependent upon the amount of fine material contained in it.

The adhesive strength of cement increases very rapidly with the fineness, at least in the early period of hardening.

The difference between coarse and fine grinding is greater in the early period of hardening than later. The fine cement hardens much more rapidly, but the coarse cement, especially in rich mortar, often reaches nearly the same ultimate strength. The attainment of extreme fineness may not therefore always be economical when the extra cost is considered.

CHAPTER IV.

THE SOUNDNESS OF CEMENT.

ART. 29. PERMANENCE OF VOLUME.

THE permanence of any structure erected by the use of cement is dependent upon the power of the cement, after the setting and hardening processes are complete, to retain its strength and form unimpaired for an indefinite period. Experiment has shown that mortars made from cement of good quality frequently continue to gain in strength and hardness through a period of several years, or at least that there is no material diminution of strength with time; and that changes of temperature, or in the degree of moisture surrounding it, produce no injurious effects upon the material. This durability in use is commonly known as the *permanence of volume* or *soundness* of the cement.

Heat has the same effect in causing expansion or contraction of cement-mortar that it has upon other materials. The coefficient of expansion for neat Portland cement, according to a series of experiments at "l'école des Ponts et Chaussées," is about the same as that of iron. For sand-mortar the coefficient is somewhat less.

When mortar which has been immersed in water is

70

transferred to dry air a slight contraction may take place in volume, together with an increase in strength, while a transference the other way may produce the opposite result; but no distortion of form or disintegration of the mortar will take place in either case if the cement be of good quality.

Sometimes cement when made into mortar sets and hardens properly, and later, when exposed to the action of the atmosphere or water, becomes distorted and cracked, or even entirely disintegrated. If the composition deviates but slightly from the normal this process of disintegration may not show itself for a considerable time, and proceeds very slowly. It thus becomes an element of considerable danger, as it is liable to escape detection in testing the cement.

The unsoundness of cement may be occasioned either by defective composition, causing the morter to yield to the action of expansions proceeding from within itself, or by exterior agencies which act upon ingredients of the cement susceptible to their influence, or are permitted to act by the method of making and using the mortar. Very little is definitely known concerning these various destructive agencies, and there is considerable doubt concerning the causes which operate in many instances. The expansive action is commonly attributed to free lime or magnesia. The exterior agencies are the action of frost, of dry air and heat, and of sea-water.

Most cements probably contain small amounts of the expansive elements, which when in small quantity act with extreme slowness, and perhaps produce no visible effect for several months after the use of the mortar; then there occurs a decrease of strength, which probably disappears with time. Cements of low index, which gain strength very rapidly in the early period, are quite apt to act in this manner, and occasionally, as already noted, the cement may not possess sufficient strength to resist, and the expansive action continues to ultimate distintegration.

The term *Permanence of Volume,* if limited to the power of the material to resist actual change of form or dimension in the body of mortar, is not necessarily synonymous with *soundness,* if by soundness we designate its power to resist disintegration over a long period. Most unsound cements fail by swelling and cracking under the action of expansives. In some cements, however, the failure occurs by a gradual softening of the mass of mortar, without appreciable change of form or dimension, the process being very slow, sometimes not noticeable for several months after the mortar is mixed.

ART. 30.   FREE LIME.

The presence of small quantities of free lime in the cement is doubtless one of the most frequent causes of disintegration in cement-mortar. The lime being distributed through the cement in small particles is hydrated very slowly after the setting of the cement, causing, through its swelling during slaking, a strong expansive force on the interior of the mortar, and producing an increase in volume, loss of strength, and perhaps final disintegration.

The effect of the lime depends upon its physical condition, and is affected by the degree of burning.

Lime burned at a high heat slakes much more slowly, and is therefore more likely to be injurious than when burned at a low temperature. Prof. Le Chatelier found that where the addition of quicklime formed by burning the carbonate produced no result, that obtained from the nitrate caused swelling and cracking in the mortar. The presence of free lime in hard-burned cements of low index is therefore of special importance, and must be carefully guarded against by securing accurate composition and complete reactions in burning.

The fineness of the cement also modifies the action of the free lime, as finely divided material will slake quicker than coarse grains, and the lime is more apt to become hydrated before setting is completed; or if the cement be exposed to the air before use, the lime in a fine state will sooner become air-slaked.

If free lime be present in such condition that it becomes slaked before the initial set of the cement, it causes no injury. If it becomes slaked during the setting or first period of hardening, the strength of the mortar may be reduced, being rendered less compact and more porous. In case this action be not sufficient to cause disintegration the loss of strength may to a great extent subsequently disappear.

When the slaking of the free lime does not take place until a longer period of time has elapsed, the danger in the use of the cement is more serious. When the expansive action becomes sufficient to overcome the tenacity of the mortar, disintegration ensues.

If the expansive action be not sufficient to overcome the tenacity of the mortar, an increase in volume and loss of strength in the mortar may take place, the extent of which is dependent upon the relative intensities of the expansive and resisting forces.

These effects may afterward gradually disappear, but they probably have the effect of making the mortar more easily attacked by external agencies. Cements of low hydraulic index, which gain strength rapidly in the early period of hardening, are particularly liable to contain appreciable quantities of free lime, which is frequently shown by a loss of strength when tests are extended over a considerable period of time.

Mortar kept under water is acted upon much more rapidly than that exposed to dry air.

ART. 31.　MAGNESIA.

Free magnesia in cement acts very much like free lime. The action of magnesia, however, is much slower than that of lime, and for this reason it is a more serious defect, because less likely to be detected in the tests applied before using. Prof. Le Chatelier mixed 5 per cent of lime and of magnesia with two samples of Portland cement, and observed the time required for the swelling to begin, resulting as follows:

	Swelling at 0° C.		Swelling in Water at 100° C.	
	Commenced.	Ended.	Commenced.	Ended.
5% lime......	in 3 hours	36 hours	immediately	in ¼ hour
5% magnesia.	in 6 monhts	....	6 hours	40 hours

In hard-burned cements of low hydraulic index any magnesia which may be present is likely to be in a free state, and hence the percentage of magnesia in such material should be low, and specifications frequently limit the amount of magnesia that may be allowed in Portland cement.

When mortar fails from this cause the expansive action may not be shown for several months after the mortar has set, and then in a comparatively short time the swelling, cracking, and disintegration take place.

Dr. Erdmenger made a large number of experiments upon the effect of adding small quantities of magnesia to Portland cement, and found * that all expanded in water and contracted in air; those containing considerable percentages disintegrated, beginning after about 90 weeks.

The fineness of the grains of magnesia, like those of lime, is important as affecting the intensity of the effect. Prof. Le Chatelier found that where magnesia coarsely ground produced swelling and cracking, the same quantity of finely ground magnesia produced swelling, but without distorting or cracking the mortar. The extent of the change in volume varies with the quantity of magnesia, increasing rapidly as the quantity is increased.

The expansion is reduced as the mortar is mixed more wet, being less as the porosity of the mortar is greater.

In light-burned cements the danger of free magnesia is greatly lessened, and many good cements of

---

* Journal Society of Chemical Industry, vol. XII. p. 927.

this class contain large percentages of magnesia. An overdose may, however, be an element of danger in any case, and several failures of mortar in Europe have been attributed to magnesia in both hard-burned and light-burned cements.

M. Durand-Claye gives an instance * where the cement in a work swelled and cracked, and an examination showed that it contained 16 to 28 per cent of magnesia. Experiments were made by mixing magnesia with good cement, and swelling resulted. The rock which served for the manufacture of the cement contained a large proportion of magnesia, which was probably present in an uncombined state in the cement. The time in which the swelling occurred was found to depend upon the amount of water available. When mixed with the normal quantity and left in dry air no expansion took place. It is therefore only dangerous in water.

Analyses óf this cement are given as follows:

Silicious Sand. $\%$	Silica. $\%$	Alumina. $\%$	Iron. $\%$	Lime. $\%$	Magnesia. $\%$	Sulphur. $\%$	Loss on Ignition.
..	14.80	8.00	4.60	47.30	24.30	0.60	0.40
..	18.30	2.95	3.60	44.80	28.15	2.30	1.90
0.35	20.70	3.35	3.65	43.30	26.70	0.15	1.80

This is not similar to the magnesian cements commonly used in the United States on account of the extremely low hydraulic index. If it be assumed that in the normal cements of this class the magnesia acts like lime in combining with silica and alumina, the presence of free magnesia might be accounted for by

---

* Annales des Ponts et Chaussées, 1886, vol. 1. p. 845.

the lack of a sufficient quantity of these hydraulic elements.

The silicates and aluminates of magnesia are known to possess, like those of lime, the property of hardening under water. Their action, however, is said to be much slower than that of the lime salts, and it has been suggested that the presence of the magnesian salts might be sometimes injurious on account of their hydrating after the hardening of the lime salts in the cement. It has not been shown, however, that any swelling takes place in the setting of these salts, and the effect may be to contribute to the final strength. Whatever the nature of the process, it is certain that some good magnesian cements continue to increase in strength over a long period, the proportionate increase in the later period being much greater than for Portland cement of even very moderate action.

ART. 32. ALUMINATE OF LIME.

The exact rôle of aluminate of lime is in many cases a matter of considerable doubt. M. Bonnami considers the basic aluminates to act as expansives in hydrating like lime, being decomposed in presence of water. The action of water upon these aluminates is very rapid, heat being given off, with the result of greatly accelerating the set of cements containing them in appreciable quantities.

The disintegration of mortar has in certain instances been attributed to the hydration of aluminates subsequent to the set. It seems probable, however, that where the aluminate is present in small quantities that its hydration usually takes place before the setting of the cement, and in fact is the first cause of the setting action.

Aluminate of lime is freely acted upon by calcium sulphate, as stated in Art. 21, forming the sulpho-aluminate ($3CaO.Al_2O_3 + 2.5SO_4Ca + 60H_2O$), which crystallizes with great expansive force. When a cement containing aluminate is exposed after setting to water containing calcium sulphate the combination of the sulphate with the aluminate may take place, with the result of causing swelling and, if sufficient in quantity, disintegration of the mortar.

Aluminates should therefore be avoided in cements to be used for mortar to be exposed to the action of sea-water, as the sulphate of magnesia acts upon the lime of the cement, forming the sulphate of lime, which then combines with the aluminate of lime, producing the expansive action. For this reason cements high in alumina are not considered desirable for marine work.

ART. 33.   SULPHUR COMPOUNDS.

The action of sulphur in cement is extremely variable, depending upon the state in which it may exist and the nature of the cement. The effect of adding sulphate of lime for the purpose of rendering the setting slower has already been discussed (Art. 21). This action depends upon the presence of aluminate of lime in sufficient quantity to take all of the sulphur into combination. When the sulphate is added in excess, or to a cement without the aluminate, it remains soluble in the mortar and is gradually dissolved out, having only the tendency to make the mortar porous.

It is pointed out by M. Candlot that if the aluminate be prevented from acting by adding slaked lime to a cement with which sulphate of lime has been mixed, the combination of the sulphate and aluminate may take place after setting, causing the destruction of the mortar through its expansive action.

The effect of the existence of sulphate in the material before burning the cement may be quite different from that of adding it afterward. In Roman cement the two seem to give analogous results; but in the heavily burned cements it may be in a state not readily soluble, and hence slow in acting upon the aluminate, thus causing the expansion to be delayed until after the set.

Mr. Spachman, who experimented upon the production of Portland cement from alkali waste, concludes * that the danger in using material containing too large proportions of sulphate of lime is due to the likelihood of forming calcium sulphide, CaS, during the burning, which afterward forms, with the iron oxide of the cement, the sulphide of iron, FeS. This upon exposure is oxidized to a sulphate of iron, changing the color of the cement to a brown, and causing it to lose much of its activity, in some instances scarcely setting at all. Mr. Spachman gives the limit of about 5% of SO_4Ca as what may be safely used.

Prof. Tetmajer states that calcium sulphate in Portland cement sometimes acts as an expansive, through

---

* Journal Society of Chemical Industry, vol. XI. p. 497.

the fact that it is readily oxidizable and expands in oxidation.

. In slag-cements the presence of calcium sulphide is thought to be less injurious. According to M. Prost, it gives a green color to the cement when kept in water, but without injury to its strength. In the air it may cause the mortar to crack.

ART. 34. EXTERIOR AGENCIES.

The principal exterior agencies which operate to cause the destruction of mortar are changes in temperature or in humidity, and the nature of the water with which it may be in contact. Exterior mechanical agencies, such as the shocks of waves or of ice and sand produced by a current, have an abrasive action and may overtax the strength of mortar in the early period of hardening, but they do not cause disintegration through injury to the cement.

The effect of frost is to set up a mechanical action through the freezing of water in the pores of the mortar and resistance to it probably depends mainly upon the strength of the mortar and its ability to resist this expansion.

The nature of the water to which the mortar is exposed is important because of the possible chemical action of salts which it may hold in solution. This is shown by the disintegration of mortar in sea-water or in sewer water which is quite sound when subjected to fresh water.

According to M. Candlot, " all hydraulic materials are alterable by pure water. A mortar traversed by pure water finally loses all coherence, the elements constituting the agglomerant being little by little decomposed. But natural water contains always carbonic acid, which intervenes in the majority of cases to arrest decomposition and close the pores of the mortar traversed by the water. When the cement does not give up its lime too readily the lime is transformed into carbonate, which forms a deposit in the voids of the mortar. If, on the contrary, the dissolving of the lime in the water which traverses the mortar is abundant, there is produced a large quantity of carbonate without cohesion, which is carried off by the water."

In regard to the effect of temperature Prof. Le Chatelier says: " At elevated temperatures certain solid hydrates lose their water and are reduced to powder, like the crystalline carbonate of soda, and cause the disintegration of the mortar. This is the case with certain aluminates of lime, especially the alumino-sulphate, but precise experiments are still necessary upon this subject.

' This dehydration occurs in dry air. This explains the well-known fact that certain cements stay for months in water and attain high strength, but the same exposed to dry hot air disintegrate into a sandy mass."

M. Candlot says that " aluminous cements are subject to alteration in surroundings exposed to alternate dryness and humidity, and also when exposed to a high temperature." It should be remarked, however, that this probably depends upon the alumina being present as basic aluminate of lime, and that cements with a high proportion of alumina, such as certain Roman cements, containing considerable sulphate, commonly give good results when used in situations exposed to changes in humidity. The Louisville cements are a prominent example of this.

When cement-mortar during the early period of hardening is exposed to very dry air, the hardening may be interfered with by the lack of moisture necessary to admit of the completion of the hydration and crystallization of the cement, thus causing a lack of cohesive strength, and perhaps ultimate destruction of the mortar. Different cements vary greatly in the extent to which they are influenced by this cause; slow-setting Portland cements being ordinarily least and the slag-cements most affected.

ART. 35.　EFFECT OF SEA-WATER.

The destructive effect of sea-water upon hydraulic mortars which are sound in fresh water is probably due to the action of magnesian salts upon the lime of the cement, thus forming sulphate and chloride of calcium. The action of these salts upon the hardening of cement-mortars has already been discussed.

When mortar in sea-water fails by swelling, the failure is usually attributed either to too large a proportion of free lime or magnesia, or to aluminate of lime in the cement. When the cement contains free lime, the expansive action is greatly intensified in sea-water as compared with that in fresh water. This may be explained by the presence of calcium chloride, which increases the rapidity of slaking of quicklime, causing the expansion to be shown sooner, and to act more violently than in fresh water. When the cement contains considerable aluminate of lime the calcium sulphate may act upon it, as indicated in Art. 32, causing the formation of the sulpho-aluminate and the corresponding expansive effect.

When mortar made from cement of good quality is exposed to the action of sea-water its durability depends largely upon the permeability of the mortar. The lime salts formed by the action of sea-water are readily soluble, and if the mass is freely permeated by water those salts may be washed out, leaving the mortar more open to the action of the disintegrating agencies. Thus mortar of any Portland cement may be injuriously affected by sea-water if used in such manner as to permit the continuous action of the magnesian salts through the mass.

The ultimate hardening of mortar in sea-water, as in fresh water, seems to depend upon the action of carbonic acid in forming a protection to prevent the operation of the elements of disintegration. When the mortar resists the penetration of the water so as to prevent its renewal in the interior of the mass, the outside soon becomes protected by the action of the carbonic acid, and effectually prevents further action of the magnesian salts.

M. Durand-Claye examined the mortar from a sea-wall where parts of it were disintegrated, and found a large proportion of magnesia, although it was not contained in the original mortar or in the portions of the wall which were still sound. The percentage of sulphuric acid was also increased in the disintegrated portions, seeming to show that the magnesia had been precipitated from the sulphate of the sea-water, and the resulting sulphate of lime had for the most part washed out.

Where the water against the wall is under pressure from one side, or where tidal flow keeps the work submerged only a part of the time, the action of the sea-water is more strongly felt than in work always entirely covered.

M. Alexander submitted blocks of cement-mortar to the filtration of both fresh and sea water.* Those in fresh water were unaffected, but those in sea-water were disintegrated in six months. Analysis showed that those in fresh water suffered a slight loss of lime and sulphuric acid, while those in sea-water were much changed by loss of lime and gain in magnesia and sulphuric acid.

M. Alexandre also found that " argillaceous or soft calcareous sand is attacked by sea-water, and mortar containing them may be decomposed although the cement is good."

---

* Annales des Ponts et Chaussées, 1890, vol. I. p. 408.

CHAPTER V.

METHODS OF TESTING CEMENT.

ART. 36. OBJECT OF TESTS.

TESTS of cement may have for their object either the examination of the quality of the material in order to determine its fitness for use, or investigation of the properties of the cement for the purpose of increasing knowledge of its behavior under the varying contingencies of use. Where experiments are made with this latter object, the tests to be applied and methods of operation must of course be dependent upon the special point to be investigated.

In many instances it may be possible to combine to a certain extent the two objects. This is particularly the case where a permanent laboratory is established to regulate the reception of material for extensive works, as in the case of the laboratories connected with the Government experiment stations in Europe. In some of these stations careful examination of every sample of cement in a number of particulars is made, with the result of accumulating a mass of valuable information regarding the characteristics of all the different kinds of cement. Systematic series of tests of this character possess much greater value as a means of

85

deducing the laws governing the action of the mortar than special examinations upon particular points, which often fail to take into account the variable nature of the material, and the necessity for exact knowledge of the nature of the cement upon which the tests are made.

The French "Commission upon the Method of Testing the Materials of Construction" recommend that in the permanent laboratories cement should be systematically tested in the following particulars: Chemical analyses; fineness; specific gravity; apparent density; homogeneity; time of setting; tensile, compressive, flexural, and adhesive strength; permanence of volume; porosity; permeability; resistance to decomposition by sea-water; and yield of mortar.

Tests of cement, as commonly made for its reception upon engineering work, have for their object only the determination of the quality of the material and its fitness for the use. Tests for this purpose must be made according to some recognized standard, and cannot closely approximate the conditions of use without impairing their value as means of judging the quality of the cement. What it is necessary to know about the cement is that it will set and harden into a solid mass, which will firmly adhere to any surface with which it may be in contact, and that it will endure through a long time without change of form or loss of solidity.

As ordinary tests must be made in a short time, but a few days at most being usually allowed for determining the quality of the material, the problem to be met in testing is to apply such tests as will enable a prediction to be made, from its behavior under them in a short time, as to what it will do in a long time under the circumstances of its use. The difficulty of this with a material varying so greatly in character and in its behavior under various conditions is evident. Having a particular brand of cement whose characteristics are known, it may readily be determined whether a given sample is of normal quality, and something may be predicted of its future from its behavior under short-time tests. Very little, however, can be done in the way of generalization, and for a new and unknown material it is only possible to state a somewhat indefinite probability as to final results.

Tests may be imposed which in nearly all cases will secure good material, but often at the expense of rejecting equally good or better material. This, however, will be unavoidable until the characteristics of the various brands of cement are more fully known, and the tests to which each should be subjected better understood.

The tests which are usually imposed to determine the quality of hydraulic cement are those of weight, fineness, time of setting, tensile strength, and soundness. Chemical analysis is sometimes made, and specific-gravity test is substituted for that of weight, or both are frequently omitted. Compression-tests are also sometimes added.

The greatest weight is usually given to the test of tensile strength, and much greater value is commonly placed upon the results of that test than they deserve. It is much the simplest and best means of making a test for strength, and is very desirable as showing the proper hardening of the mortar, but cements cannot be graded in value by the strength attained in a short time. Cement giving high early strength is to be relied upon only in so far as it has been shown by experience capable of subsequently maintaining such strength. The attempt to produce cement which will develop great strength on short-time tests is liable to result in lowering the hydraulic index, or the addition of calcium sulphate, and sometimes in the presence of free lime, giving a material more likely to be unsound than one of more moderate strength.

The test for soundness or permanence of volume is an important one, as giving an indication of the probable durability of the material; but in this, as in the other tests, a knowledge of the usual action of the material will contribute greatly to the proper interpretation of the test.

The test for fineness is also important as bearing upon the power of the cement to take sand.

It was recommended by the committee of the American Society of Civil Engineers upon a uniform system of testing, that tests for quality be limited to the above three most important tests—fineness, tensile strength, and soundness; and this recommendation is now commonly followed in the United States, although the test for soundness as usually made is of little value.

ART. 37.   APPARENT DENSITY.

The *apparent density* of cement is measured by determining the weight of a given volume of the material. This test is made as a means of showing whether the process of manufacture has been well conducted. If the cement be not thoroughly burned or if it lack homogeneity so that in portions of it the combinations are not complete, the weight is less than when the material is homogeneous and well burned. Variations of composition also affect the weight, so that there may be considerable variations in the weight of various cements of good quality, equally well burned.

The apparent density is affected by the fineness to which the cement is ground; the coarser the particles of the cement, the greater its weight per unit volume. The weight test when employed should therefore be combined with one for fineness to prevent the attainment of heavy weight by coarse grinding.

The test for apparent density is not usually employed for the reception of material, as it is somewhat indefinite in result. It is, however, sometimes included in specifications in England, and is used in many European laboratories where a careful study is made of the properties of cement.

As the cement powder may be packed in the measure so as to give very different weights for the same volume, it is necessary to use a uniform method of filling the measure in determining the weight. The common method of conducting the test is to pass the powder through a sieve and allow it to fall through a funnel or down an inclined plane through a given height into a measure, which when full is struck and weighed. The height of fall and the size of the measure both affect the result, the cement packing closer in a large than in a small measure.

In Europe several appliances are used for testing apparent density.

Tetmajer's Apparatus. — The apparatus of Prof. Tetmajer is used in a number of the leading laboratories. It is shown in Fig. 1, and consists essentially

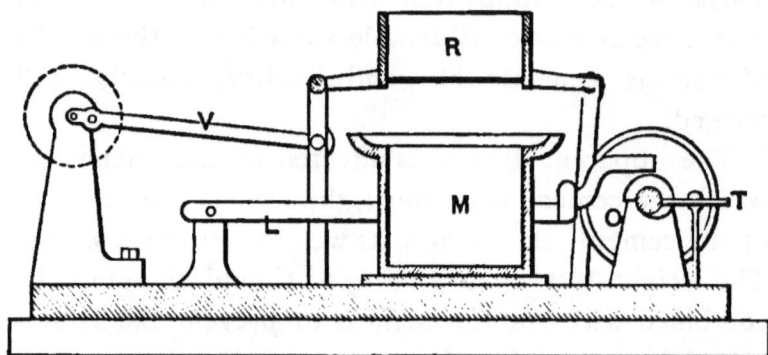

FIG. 1.

of a cylindrical measure (*M*) of 1 liter capacity and 10 centimeters high, provided with ears which catch upon a frame formed of two levers (*L*). The frame is raised and dropped at each turn of the hand-wheel by the cam (*O*), thus giving a succession of jars to the measure.

Above the measure a sieve (*R*) is oscillated upon a system of levers which are hinged to the base, and moved by the rod (*V*), giving two oscillations at each turn of the hand-wheel. The number of revolutions is recorded by the revolution-counter (*T*).

In the operation of the apparatus the cement is filled into the sieve and shaken through by the oscillations produced by turning the hand-wheel. It is caught in the measure and jarred down by the raising and dropping of the frame. About 500 revolutions

are necessary to secure the best results in compacting
the powder in the measure. The compactness is found
to vary with the rapidity of motion, a moderate speed
of about 200 revolutions per minute giving a maximum
effect, and being considered most desirable.

Inclined-plane Apparatus.—The inclined-plane apparatus for apparent density has been used in a number of forms, one of which, employed in France and
recommended by the " Commission des Méthodes
d'Essai des Matériaux de Construction," is represented in Fig. 2.

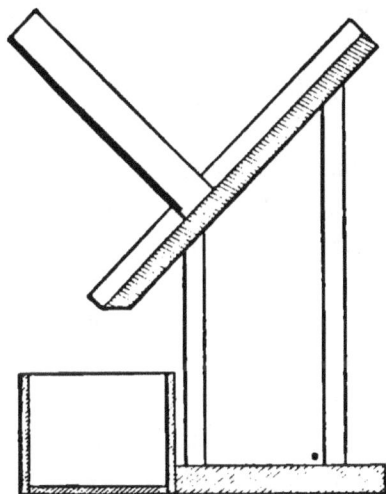

FIG. 2

The inclined plane is formed of sheet zinc 30 cm.
long and inclined at 45° with the horizontal. It is 10
cm. wide for the upper two thirds of its length, and
through the lower third diminishes gradually to 5 cm.
at the lower end. The zinc is turned up at the sides
to form a channel in which the material may slide.

At the point where the larger plane begins to narrow a second sheet of zinc, 20 cm. long and 10 cm. wide, is set at right angles to the first, leaving an opening of 1 cm. The measure 10 cm. high and of 1 litre capacity is placed with its top 5 cm. below the lower edge of the plane.

The cement is poured in small quantities on the summit of the secondary plane so slowly as not to clog the opening between the planes, until the measure is full when it is struck and weighed.

A single inclined plane of somewhat greater length (50 cm.) is sometimes used, the cement being sifted upon the upper end and allowed to slide directly into the measure. It is said, however, to give less uniform results than the double plane unless handled with extreme care.

German Funnel Apparatus.—This apparatus was recommended by the German conference upon methods of testing materials, as was also the Tetmajer apparatus.

The funnel is formed of a hollow cone with its axis vertical, as shown in Fig. 3. The height of the cone is 18 cm., its upper base is 20 cm. and lower base 2 cm. in diameter, terminated at the lower end by a second cone 5 cm. high, with a lower base 1.6 cm. in diameter. The funnel is supported upon a tripod, with its lower end 20 cm. above the table and 10 cm. above the top of the liter measure. To facilitate the flow of cement into the measure a rod 7/10 cm. in diameter is rotated in the axis of the funnel. This rod is guided by two cross-rods supported upon the interior surface of the funnel.

In the operation of this apparatus sufficient cement
to fill the measure is placed in the funnel, and the rod
is then rotated, about 45 revolutions per minute, by
gear or by hand until the material has passed through

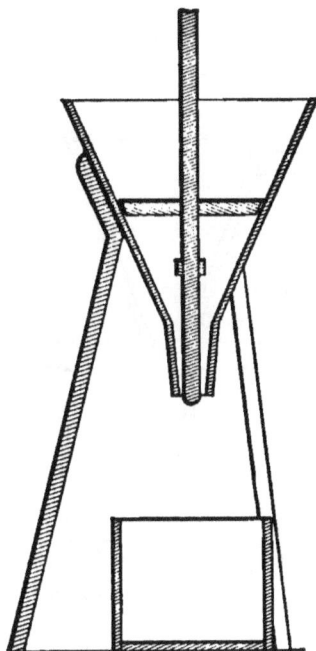

FIG. 3.

and filled the measure, which is then struck and
weighed.

Sieve and Funnel Apparatus.—This apparatus as
used in France is shown in Fig. 4. It consists of a
funnel with a sieve fitting into the upper part of it
and the measure below. The cement is put into the
sieve, and gradually worked through by the use of a
spatula. It then slides down the funnel into the
measure until that is filled.

The "Commission des Méthodes d'Essai des Matériaux de Construction" made a careful comparison of
results obtained by the various methods.   They found
the German funnel apparatus quite precise in its

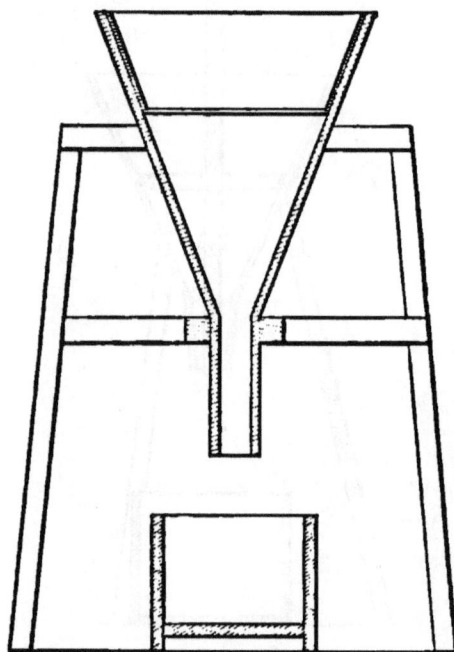

FIG. 4.

results with certain materials, but that with some
cement it always became clogged by the packing of
the material in the funnel.   The Tetmajer apparatus
is capable of great precision, but is somewhat complicated, and requires careful manipulation to secure
always the same rate of filling the measure and the
same amount of compacting.   The inclined plane and
the sieve and funnel apparatus are found to give good
results, and are recommended by the commission for

use in France. The two latter give nearly identical results, the German apparatus somewhat higher and the Tetmajer apparatus much larger results than the others.

The following method for apparent density was recommended in the preliminary report of the committee of the American Society of Civil Engineers upon cement testing, but has never come into common use:

"Procure a cylinder of a height of 6 inches, having an interior area of 2 square inches. Sifting the cement to be measured so that it may not be compact, weigh carefully 5 ounces if of Portland cement and 4 ounces if of natural cement, and pour the same into the cylinder, which should stand upright with its lower end resting upon a close-fitting and suitable base; then, without shock or sudden impact, lower a close-fitting piston, moving without friction, slowly down the cylinder on to the cement; said piston and its attachments to weigh 50 pounds. After resting thereon one minute, remove the same and ascertain the bulk of cement thus compressed."

In making tests for apparent density it is advisable to sift the cement, and use only that portion which passes the finest sieve, thus making the result to a certain extent independent of the fineness of grinding. To accomplish this the sieve used should be as fine as possible in order to eliminate all but the impalpable powder. In Europe a sieve of 5000 meshes per square centimeter is employed for this purpose, corresponding to the No. 180 sieve, 32,400 meshes per square inch.

The ordinary weight of Portland cement varies from 70 to 100 lbs. per cubic foot, depending largely upon the method of making the tests. Natural cement is usually somewhat lighter.

ART. 38.   SPECIFIC GRAVITY.

The determination of *specific gravity* is often substituted for that of apparent density, and is a much better guide to a knowledge of the actual density of the material, as it is not subject to the fluctuations due to fineness or method of determination which characterize the weight tests. The differences of specific gravity to be determined are, however, very small, and great care is necessary in the manipulation of the test in order to obtain reliable results.

The test for specific gravity is commonly made by immersing a known weight of the cement in a liquid which will not act upon it, and obtaining its volume through noting the volume of liquid displaced. In making the test by this method it is necessary that all the air-bubbles contained in the cement powder be eliminated, and that the volume obtained be that of the cement particles only.

Schumann Volumenometer.—Several forms of apparatus have been used for this purpose. Of these the Schumann volumenometer, shown in Fig. 5, is perhaps the most common. It consists of a graduated tube, the bottom of which is ground to fit closely into the top of a flask.

In the use of the apparatus the tube is placed upon the flask and filled with benzine to the zero-point on the scale. 100 grammes of cement are then weighed and carefully poured into the top of the tube so as to sift gradually through the liquid, thus allowing the air to escape.

The elevation of the surface of the liquid in the tube

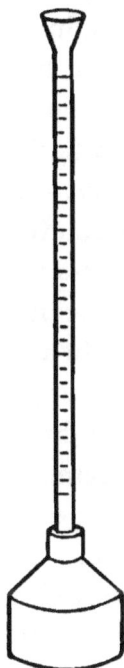

FIG. 5.

gives the volume of the cement. The scale as ordinarily made has a range of 40 cubic centimeters, and is graduated to 1/10 centimeter. This volumenometer gives very satisfactory results when carefully used, but much care is required to fully eliminate the air and prevent the powder from adhering to the surface of the tube. It is well in operating in this manner to introduce the liquid through a long funnel, thus keeping the inner surface of the tube dry.

In some instances greater accuracy may perhaps be obtained by filling the apparatus to the zero-point with liquid and weighing the whole, then pouring in enough cement through a funnel to raise the surface of the liquid through a definite volume, and determining the weight of cement by reweighing the whole. This eliminates the error due to loss of cement powder in introducing it into the flask.

Candlot's Volumenometer.—M. Candlot has modified the Schumann volumenometer by closing the upper end of the tube with a glass bulb. In using this form of the apparatus the tube is turned bottom upward and the bulb filled with benzine. The flask is then placed upon the tube, the apparatus is inverted, and a reading taken of the height reached upon the tube. A weighed quantity of cement is then placed in the tube, which is turned upside down and separated from the flask for the purpose. The flask being replaced and the whole again inverted, the volume is shown by the use of the liquid in the tube. In order to eliminate the air-bubbles the apparatus is shaken before taking the second reading so as to thoroughly mix the liquid and the cement.

This method is much more rapid than that with the Schumann volumenometer, but in practice is hardly so satisfactory in result.

Le Chatelier's Volumenometer.—This apparatus is shown in Fig. 6. It consists of a flask the top of which is drawn out into a tube 1 cm. in diameter and 20 cm. long. Above the middle the tube enlarges into a bulb for a short space, and then again continues with uniform diameter to the top. The flask to a point marked on the tube just below the bulb has a capacity of 100 cubic centimeters. From this point to a mark above the bulb the capacity is 20 cubic

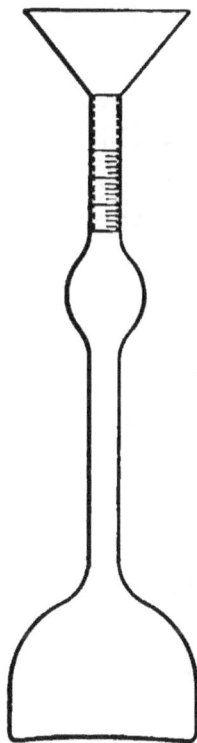

FIG. 6.

centimeters. This latter mark is very carefully determined, and the upper part of the tube is graduated to 1/10 cubic centimeter.

In using this apparatus, the flask is filled with liquid to the mark below the bulb, and the cement is then slowly introduced through a funnel, and settles through the liquid into the flask, the air being eliminated by

its long passage through the liquid. Cement is
added until the surface of the liquid rises to the 20-
cubic-centimeter mark. The weight of this volume
of cement is obtained by weighing the apparatus before
and after the cement is introduced. Or, this volumenometer may be used in the same manner as that
of Dr. Schumann, the bulb serving to prevent the
cement sticking to the sides of the tube.

Mann Gravimeter.—This apparatus consists of a
flask which when filled to a certain mark upon its neck
contains an accurately known quantity of liquid. A
graduated tube with a stop-cock at its lower end contains when full the same quantity as the flask.

A weighed quantity of cement is placed in the flask
and the tube is filled with the liquid to be used. The
liquid is then allowed to run from the tube into the
flask, reliance being placed upon shaking the flask to
eliminate air-bubbles, until the flask is filled to the
volume mark. The volume of liquid remaining in the
graduated tube is then equal to the volume of cement
powder. The complete eradication of the air is a
matter of difficulty, and if the operation is conducted
altogether in the air the change of temperature may
be sufficient to affect the result. This method in
practice is not likely to give very concordant results.

Erdmenger's Volumenometer is a modification of the
Mann apparatus, looking to the maintenance of a constant temperature during the test. This arrangement
is shown in Fig. 7. It consists of a graduated tube
of 50 cubic centimeters capacity, partially enclosed in
a larger vessel which acts as a cooler. The upper end
of the tube is closed with a ground-glass stopper, the

lower end with a stop-cock.   Near its lower end the
tube has a horizontal branch, also closed by a cock,
and connecting it with a jar of 1 liter capacity having
two openings at the top, one being connected with the
measuring tube, the other with a rubber pressure-ball.
The cooler has two openings at the top, one of which
serves to permit the passage of air, the other to admit
a small thermometer.   To fill the cooler with water,

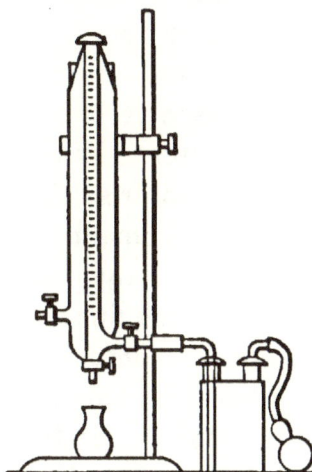

FIG. 7.

or empty it, an opening is placed at the bottom which
may be controlled by a stop-cock.   A narrow-necked
flask of the capacity of 50 cubic centimeters is used for
the measurement of the volume of the cement.

In making the test the apparatus is kept at a temperature of about 60° Fahr. by filling the cooler with
water at that temperature and standing the double-
necked jar containing the liquid to be used in the test
and the measuring-flask in vessels of water of the same
temperature.   The measuring-flask is then filled to

the 50-cubic-centimeter mark with the fluid, and the test is made as with the Mann apparatus; or the mixing of the liquid with the cement may be done by putting the cement in a small funnel, and washing it through with the liquid, thus eliminating the air.

Any of these volumenometers may give good results when carefully handled, the Schumann and Le Chatelier forms being easiest to handle satisfactorily.

Benzine or turpentine is usually employed as the fluid, and it is important that the liquid should be maintained at a nearly uniform temperature during the test.   For this reason it is common to immerse the apparatus in cool water.   Where benzine is used the temperature should not rise above 60° Fahr.   It is desirable also to sift the cement to be tested through a fine sieve, on account of the better elimination of air-bubbles possible with the fine material.

In order to make the determination of specific gravity of value, it must be reliable to two decimal places.   Portland cement varies from about 3.00 to 3.18, and is usually above 3.05.   Natural cements vary from 2.75 to 3.05.   An inferior limit is sometimes fixed in specifications—usually for Portland cement about 3.00 or 3.05, and for Roman cement about 2.80.

The presence of the volatile elements due to incomplete burning, or of adulterations added after the burning, tends to lower the specific gravity.   The quantity of adulteration, however, needs to be considerable before it becomes appreciable in the results of this test.

The specific gravity, unlike the apparent density, is not affected by the fineness of grinding; and it has been suggested by examination of the results of certain experiments that a comparison of the two tests for the same material may sometimes give a better determination of the actual fineness than can be obtained by the use of sieves. This fineness is shown practically by the greater ability of the fine cement to " take sand " without losing in early strength.

ART. 39. TESTS FOR FINENESS.

The fineness to which a cement is ground is usually considered a matter of importance, as upon it depends very greatly the early adhesive strength of the mortar and the ability of the cement to take sand.

A test for fineness is nearly always included in specifications for cement, and the test is particularly necessary where the tensile strength is tested for neat cement only. In such case the attainment of a proper strength neat, together with a fair degree of fineness, practically insures that the cement will give good results when used with sand.

The fineness which should be required is largely a matter of relative economy; the finer the cement, the larger the quantity of sand that may legitimately be used with it. The coarse parts of the cement are to be considered as inert material, or practically as a certain amount of sand already mixed with the cement. It is a question therefore of relative costs of different degrees of fineness.

There is, however, some dispute as to the value of fineness. Some European authorities question the wisdom of a fineness test. It is well known that the effect of fineness in the strength of sand-mortar disappears to some extent with time, but the impalpable powder seems to be the really valuable part of the cement, and if this be omitted the cement loses its value.

Prof. Le Chatelier in his microscopic examination of mortar found that, after setting, in the more fine particles no trace is left of the grains of the cement. With the larger ones the central part of the grain remains unaltered. It seems that the grains which are completely attacked are limited to 0.1 millimeter in diameter, but further study is needed upon this point.

Coarse grinding also, as has been elsewhere noted, increases the intensity of the action of expansives which may be contained in the cement, causing a coarse-ground cement to expand and crack, when perhaps if finely ground it would be unaffected.

The test for fineness simply consists in sifting the cement through a sieve or a set of sieves, and observing the amount retained by each sieve.

The committee of the American Society of Civil Engineers upon "standard tests" recommend the use of sieves of 2500, 5476, and 10,000 meshes per square inch. Specifications usually, however, require only a single sieve—generally that of 2500 meshes, but sometimes that of 10,000 meshes. A more general use of the finer sieve would undoubtedly be advantageous, as it is now generally admitted that all material coarser than that dimension is practically inert, and a real measure of useful fineness is not given by the 2500-mesh sieve   A common requirement is that not more than 10% by weight of the cement be retained upon a sieve of 2500 meshes, or that not more than 20% be retained upon that of 10,000 meshes, or both. Most of the cements commonly in use in the United States easily comply with these requirements. Very many of them do not give a residue on the coarse sieve of more than 1% to 3%, or on the fine one of more than 8% to 10%. Some cements, however, seem to be bolted with special reference to passing the test of the 2500-mesh sieve, and are very coarse when tested with a finer one.

The sizes of wire of which the sieves are made is of course important as regulating the sizes of the openings, and should always be stated; the common standard is that the diameter of wire should be about 1/3 of the spacing between centres. The sieves of the American Society of Civil Engineers mentioned above are of Nos. 35, 37, and 40 wire-gauge.

It is not usually practicable to get sieves with perfect regularity either of spacing or diameter of wires. A sufficiently near approximation for practical work may be obtained by using care in selecting the sieve, but the gauge frequently offered for this use differs very considerably in the sizes of openings for the same number per inch, and sometimes the openings are quite irregular in size in different parts of the same sieve.

ART. 40.   RATE OF SETTING.

The rate of setting of cement is tested for the purpose of determining if it be suitable for a given use, and not as a measure of the quality of the material. For most purposes, where immediate setting is not required to prevent disturbance of the mortar before hardening, the moderately slow-setting cements are found most convenient, as they need not be handled so quickly, and may be mixed in somewhat larger quantities.

Testing for time of setting consists in arbitrarily fixing two points in the process of consolidation, which are called the beginning and the end of setting. These points are differently determined in the various methods of testing, and are not marked by any distinguishing phenomena which admit of definite determination.

The method recommended by the committee of the American Society of Civil Engineers is that proposed by General Gillmore, and consists in mixing cakes of neat cement, about 2 or 3 inches in diameter and 1/2 inch thick, to a stiff plastic consistency, observing the time when they will bear a needle 1/12 inch in diameter sustaining a weight of 1/4 pound, and noting this as the beginning of setting; then continuing the observations with a needle 1/24 inch in diameter carrying a weight of one pound until the material is sufficiently firm to bear this, when it may be called fully set.   The committee call those cements which set in 1/2 hour or less, quick-setting; those requiring more time, slow-setting.

The time of setting is often roughly determined in practice by making small cakes of mortar and observing when they will resist penetration under a light pressure of the thumb-nail. This is a standard test in Germany.

For ordinary practical purposes these methods are sufficiently accurate, as all that is necessary is to know whether the cement sets quickly or slowly, but for experimental and comparative purposes more elaborate methods are valuable. The beginning of setting is the point of most value, as the cement in practice should be used before that point is reached, in order that it may not be disturbed after the stiffening has begun.

In Germany and France the Vicat needle is commonly employed for accurate determinations. This arrangement is shown in Fig. 8. By this method a

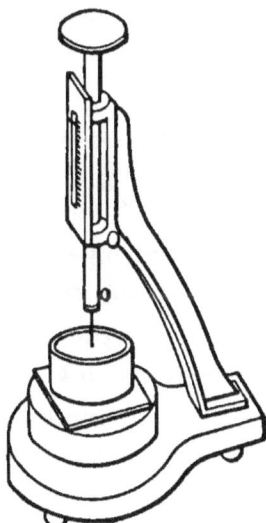

FIG. 8.

briquette of neat cement is made in a cylindrical brass or rubber mould 10 centimeters in diameter and 4 centimeters high, placed upon a plate of glass or metal, the cement being mixed to a plastic consist-

ency as determined by the consistency test. The
apparatus is so arranged that a weight of 300 grammes
may be brought either upon a needle of 1 square millimeter section, or upon a cylindrical plunger 1 centimeter in diameter, and allowed to settle into the
cement, the depth of penetration being shown by a
scale along which the weight slides. As soon as the
mould is filled with the mortar it is placed in the
apparatus, and the plunger, sustaining the 300
grammes, is brought to the surface of the briquette
and allowed to sink into it. If the plunger penetrates
to a point 6 millimeters from the bottom the mortar
is of proper consistency for the test. The needle is
then substituted for the plunger, and the time when
the needle first refuses to sink entirely through the
mortar is observed and noted as the beginning of setting; the time when the needle first rests upon the
briquette without penetrating it is considered the end
of setting.

The accurate determination by this method of the
points where the set is said to begin and end is a
matter of some difficulty, as the lack of perfect homogeneity causes the needle to sink more deeply in some
parts than in others, and the cement sets more rapidly
at the circumference than in the interior of the mass.
However, these defects are not very serious when due
care is exercised in mixing the mortar, and the penetration is not taken too near the edges. The time of
completion of set is much less well defined than that
of beginning of set, as there is usually a considerable
period during which a very slight penetration takes
place, decreasing insensibly to final disappearance.

The point to be used for completion of the set is that
at which the penetration becomes very small, so that
the curve of penetrations becomes practically horizontal.   Such a point is usually fairly well defined.

Various modifications of this method have been proposed for the purpose of securing greater uniformity
in result, but they have not come into general use.
M. Bonnami has proposed to modify the test by varying the weight upon the needle instead of the depth
of penetration.   He measures the time at which
various weights will cause the penetration of the needle
to mid-depth in the mortar, beginning with 50 grammes
which marks the beginning of setting, and increasing
to 3000 grammes, which gives the end of setting, thus
obtaining a curve of times in terms of weights sus-
tained.   The penetration to mid-depth is selected as
the point of maximum variation.

The time of setting is usually tested upon paste of
neat cement, on account of the difficulty of obtaining a
satisfactory test with sand-mortar.   The Vicat needle
is quite useless when sand is employed because of the
interference of the grains of sand with the descent of
the needle.   Rough tests of sand-mortar by the
ordinary methods may readily be made with sufficient
accuracy for practical purposes, and are very desirable
as showing more nearly what may be expected of the
mortar when used.   The rate of setting of neat mortar
gives but little indication of what the action may be
with sand.   For different cements a mortar of 3 parts
sand to 1 of cement may require from about $1\frac{1}{2}$ to 8
or 10 times as long as neat paste when the same sand
and method of mixing are employed.

Several propositions have been made with reference to a standard test for the purpose of comparing the rate of setting of sand-mortars. These usually have been to substitute cylinders of larger diameter for the Vicat needle, in order to reduce the effect of the sand grains, and then to use correspondingly heavy weights to produce penetration. The usual method is to determine the weight necessary to indent the surface of the mortar. Thus in one apparatus a cylinder 1 cm. in diameter is employed: when the mortar will just bear a weight of 400 grammes it is considered as beginning to set; when it will sustain 10,000 grammes the setting is complete.

M. Feret has also proposed to make standard tests by using fine sand composed of grains which pass the sieve of 75 meshes and are held by one of 180 meshes per linear inch, the test being made as for neat cement with the Vicat needle. This serves as a comparison of the effects of sand upon different cements.

In making tests for rate of setting, the temperature of the ingredients of the mortar before gauging, that of the atmosphere in which it is gauged, and of the air or water in which it is placed during setting have a very large influence upon the results. A temperature of 60° to 65° Fahr. is usually accepted as standard, although the air in the laboratory may have a somewhat higher temperature—perhaps 65° to 70°.

The amount of water used in gauging and the methods of mixing the mortar and filling the mould are important. For the purpose of regulating these, the consistency of the mortar is prescribed. This is probably the best means of arriving at uniform results, but it should be pointed out that the same consistency may be arrived at in two ways—by using a small quantity of water and working thoroughly, or by using a larger quantity and working less. The quantity of water which will bring the mortar to proper consistency after three or four minutes of vigorous working is the most desirable, but there is opportunity for considerable variation in the results of the test as carried out by different operators.

A number of automatic appliances have been devised for the purpose of making these tests without constant attendance. That of Professor Fuertes, in the laboratory of the College of Civil Engineering, Cornell University, consists of a trough $1\frac{1}{2}$ inches wide by $2\frac{1}{2}$ inches deep, in which the mortar is placed. A car is drawn along a track over the trough by means of a screw rotated by a clock, thus giving a motion varying uniformly with time. From the car is suspended a Vicat needle, which is dropped and raised at regular intervals by the action of a small stream of water. A pencil attached to the shaft of the needle draws the curve of penetrations upon a board at the side of the trough.

ART. 41. CHANGE OF TEMPERATURE DURING SETTING.

Observations of the change of temperature during setting are commonly taken in many of the European laboratories. It has been thought by some observers that the points of beginning and end of setting might be more accurately marked by observing the change of temperature than by the needle test. The operation of setting is a chemical action which takes place with the disengagement of heat, but in many cases the rise in temperature is so slight and indefinitely marked that it would be difficult to use it in this manner.   The amount of the variation in temperature varies somewhat with the activity of the cement, increasing rapidly as the cement sets more quickly. The total rise of a quick-setting cement may reach 15° or 20°, while in a very slow one it may be quite imperceptible.

The change in temperature also varies with the nature of the cement, and attempts have been made to connect it with the soundness of the material, particularly the presence of free lime.   This, however, does not seem to be supported by facts, or at least the indications are very indefinite.   Expansives which are slow in action, and therefore dangerous in the cement, are not likely to cause increase in temperature during setting.

The test for change in temperature is ordinarily made by placing the mortar in a cylindrical mould, like that used with the Vicat needle, fitted with a cover through which is an opening to permit the introduction of the thermometer.   This cover prevents the mortar coming into contact with the air or water in which it may be placed, thus neutralizing the effect of the internal change of temperature.

Some very slow-setting cements show a fall in temperature if left exposed to the air while setting, probably due to surface evaporation.

This test does not seem of importance as a measure of the quality of cement, but it is worthy of attention in a systematic study of the properties of the material, and may be capable of giving interesting results.

CHAPTER VI.

TESTS OF THE STRENGTH OF MORTAR.

ART. 42. METHODS EMPLOYED.

THE strength of mortar is frequently tested in three ways: the tensile test is the one more commonly employed, but compressive and transverse tests are also often used.

The test for tensile strength is made by making briquettes of the mortar in moulds having a definite section at the middle,—in the United States usually one inch square,—and enlarging at the ends to fit in clips by which they may be placed in the testing-machine and pulled apart by direct tension. This test is in common use, because it can be more readily and uniformly applied than the others, and seems, when coupled with other tests, to give a satisfactory indication of the value of the material.

The compressive test consists in crushing small blocks of the mortar between the jaws of the testing-machine and weighing the force required. This test is more difficult in manipulation to secure uniform results, and also requires much heavier appliances, on account of the high resistance offered by the material to crushing.

The transverse test is made by moulding the mortar into bars; the bar is afterward placed horizontally upon supports near its ends, and broken by a load brought upon its middle, causing it to break by bending.

The proper conduct of any test for strength is a matter requiring care and experience. There are many points connected with the circumstances and manipulation of the work which have an important bearing upon the result. These are: the form of the briquette; the method of mixing and moulding; the amount of water used in tempering the mortar; the surroundings in which the mortar is kept during the hardening; the rate and manner of applying the stress; the temperature at which all of the operations are performed. In order to secure uniform results it is essential that the tests be standardized in all of these particulars. Much has been accomplished in this direction during recent years, but there is still great disparity in the results of different operators, undoubtedly due mainly to differences in making the briquettes.

Every laboratory seems to have to a certain extent its own practice, which makes its work incomparable with that of any other laboratory. Even where presumably the same methods are used it is very difficult to frame rules that all will understand alike, while in all cases the personal equation of the operator is an important matter in hand-work.

The committee upon standard tests of the American Society of Civil Engineers in their report call attention to this matter in the following words:

" The testing of cement is not so simple a process as it is sometimes thought to be. No small degree of experience is necessary before one can manipulate the materials so as to obtain even approximately accurate results.

" The first tests of inexperienced though intelligent and careful persons are usually very contradictory and inaccurate, and no amount of experience can eliminate the variations introduced by the personal equation of the most conscientious observers. Many things, apparently of minor importance, exert such a marked influence upon the results, that it is only by the greatest care in every particular, aided by experience and intelligence, that trustworthy tests can be made."

Experience, since the report of the committee was made, has shown that the difficulties in the way of uniformity in such tests are much greater than was then imagined.

The variations in the results of the tensile test between the work of different experienced operators working by the same method and upon the same material are frequently very large, and often make all the difference between the acceptance and rejection of the cement. Differences of 40% to 60% with neat cement are not uncommon, while for sand-mortar they are much greater.

An investigation of this matter by Prof. J. M. Porter, of Lafayette College, is interesting in this connection. He divided a sample of cement into a number of parts, sending each to a different laboratory with the request that tests be made of it in 1 to 3 mortar, according to the rules recommended by the committee of the American Society of Civil Engineers. The resulting average strengths of each of the nine laboratories were as follows, in pounds per square inch: 75, 102, 114, 133 and 140, 153, 163, 176, 225, 247. These results (see *Engineering News*, March 5, 1896) show that the lowest strength was but 30% of the highest, while the remainder were quite evenly distributed between the two extremes. Each result was the average of five briquettes, which agree fairly well among themselves.

If the results of experienced men in the permanent laboratories vary so much, what is to be expected of tests made by less experienced men for the reception of material upon temporary work, and how can a specification be framed which shall fairly determine the value of the material? Evidently, to secure proper results with hand-work, the inspector must first be calibrated, and the specifications drawn in accordance with the practice of the laboratory. It is at least very desirable that some means be devised by which the work of these tests may be made automatically, and the personal factor eliminated in so far as possible.

In standard tests it is customary to adopt a nearly constant temperature of 60° to 65° Fahr. for the air in the laboratory in which the briquettes are prepared and the tests made, and about the same or slightly less for the water used in tempering and that in which the mortar is immersed during hardening.

ART. 43.   FORM OF BRIQUETTE.

Briquettes of mortar for tests of strength are commonly formed in moulds of metal of the form to be used in the tests.   As the size and shape of the specimens have an important effect upon the result, it is necessary to adopt standard dimensions in order to obtain uniform results.

For compressive tests a parallelopiped, usually a cube, is employed.   In the United States a cube whose edges are each two inches in length is commonly used, although sometimes an inch cube is used. In Europe generally the standard specimen is a cube with edges seven centimeters in length.

The French '' Commission upon Methods of Testing Materials,'' however, rejected the rectangular section for compression specimens on the ground that it is difficult to so fill the corners of the moulds as to make homogeneous briquettes, and they recommend the cylindrical form as preferable.   They also recommend the use of half-briquettes obtained by the tension test as blocks for the crushing test.

For transverse tests bars of rectangular section are used.   Different experimenters have used quite different dimensions, and there is no size which may reasonably be called a standard.   Those who have proposed the adoption of this test in place of that for tension in the acceptance of material have usually advocated a test-piece of a section one inch square and from eight to twelve inches long, although sometimes the section is made two inches square.

For tensile tests many forms of briquettes have been tried, but at present there are but two in common use: the one recommended by the committee of the American Society of Civil Engineers (shown in Fig. 9), which was derived from that used by Mr. Grant in England, is now the standard in the United States, and commonly used in England; the other is the form adopted by the Association of German

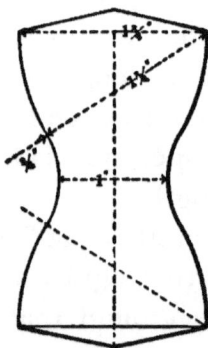 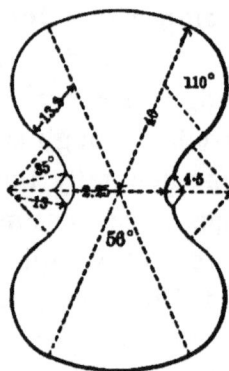

FIG. 9.   FIG 10.

Cement Makers, and is the standard in Germany, and generally employed in France. This form is shown in Fig. 10, which gives the dimensions in millimeters. The middle section is 22.5 mm. wide by 22.2 mm. thick, giving a cross-section of 5 square centimeters.

Comparative tests of briquettes of the two forms indicate that the English form gives higher results than the German, the difference being commonly for neat briquettes as much as 30% to 40% of the smaller. This may perhaps be accounted for by the fact, stated by Mr. Faija, that a sudden change of cross-section is always an element of weakness, and while the English form diminishes gradually from the ends to the middle, in the German form the area is suddenly decreased by a circular notch at the middle.

It may also be noted that the surface upon which the clip catches the briquette when being tested is in the German form inclined at a greater angle to the centre line of the briquette, and consequently the adjustment in the clips to produce axial stress is less perfect.

In England and the United States the standard area for the middle is one square inch; in Germany and France a little smaller, being, as already noted, five square centimeters. · The use of these small sections is advantageous, as it admits of lighter apparatus in making the test, and because greater uniformity is easily attainable in making the briquette. The work also is facilitated by the fact that less mortar is required for each specimen than with larger sections, so that more briquettes may be prepared from each mixing.

The size of the breaking section has an important effect upon the strength, the smaller sections giving much higher strength per unit of area than the larger ones. Thus for neat cement a change from a section 1 inch square to one 2 inches square has been found to lessen the tensile strength per square inch to about one half that of the smaller section. M. Durand-Claye has shown that the strength varies more nearly with the perimeter of the section than with its area, and that the interior may be removed without loss of strength.

M. Alexandre made a number of experiments upon

the relative strengths of briquettes of different sizes.*
He found that large briquettes gave much less strength
per unit area than small ones, but for sand-mortars
the effect diminished as the proportion of sand increased, and the difference also became less as the
age of the mortar increased, seeming to indicate that
the effect may be partly due to the more perfect
hardening of the small specimens

ART. 44.   QUANTITY OF WATER USED IN GAUGING.

The determination of the proper consistency for
mortar is very important in its effect upon the results
of tests of strength.   The effects upon setting and
hardening of varying the quantity of water used in
mixing have already been discussed, Art. 24.

In making standard tests it is common to regulate
the quantity of water by trying to bring the mortar
to a normal consistency which shall be uniform for all
tests.   Different cements require very different quantities of water to reach the same consistency, and in
the use of sand-mortar the nature and condition of
the sand may also cause considerable variation.   It
should also be noted that the consistency of mortar
does not depend altogether upon the quantity of water
used, but may be varied by the manner and extent of
working the mortar during the gaugings.   In French
and German practice it is required that the mortar be
vigorously worked for five minutes.

Under the different systems of making briquettes
there are two consistencies employed as standards—

---

* Annales des Ponts et Chaussées, 1890, vol. II. p. 277.

the plastic and the dry.   In using each of these methods the quantity of water used and the consistency
reached vary greatly in different places, as the man
doing the work may interpret the terms used in describ-
ing the desired state of the mortar.

The rules recommended by the committee of the
American Society of Civil Engineers describes the
condition of the mortar to be used as " stiff and plastic," thus leaving much to individual judgment.

In Europe two methods are employed for determin-
ing the proper consistency of plastic mortar.   The
Tetmajer method consists in determining the normal
consistency by the method already described for find-
ing the time of setting by the Vicat needle (Art. 40).
This method was recommended by the conference of
German, Austrian, Swiss, and Russian engineers at
Dresden, and also by the French commission.

The Boulogne method is commonly used in France,
and also approved by the French commission.   It is
as follows:

The mortar is to be vigorously worked for five
minutes to bring it to the required consistency.

" 1. The consistency of the mortar should not
change sensibly if the mixing be continued three
minutes after the expiration of the required five
minutes.

" 2. If a small quantity of the mortar be taken up
on the trowel and allowed to fall upon the mixing slab
from a height of 50 centimeters it should be detached
from the trowel without leaving any small particles
adhering, and after falling should approximately retain
its form without cracking.

"3. A small quantity taken in the hand and patted into a round form, until water flushes to the surface, should not stick to the hand, and, when allowed to fall from a height of one-half meter, the ball should retain its rounded form without showing any cracks."

To meet these requirements leaves but a narrow limit within which the consistency may vary. If a slightly too small quantity of water be used, the mortar would crack upon falling. If the quantity be very slightly too great, the mortar continues to soften upon further working, becomes sticky, and loses its form upon falling.

Mortar of the dry consistency is used when the briquette is to be made by beating the mortar into the moulds. For this purpose the mortar is required to have the consistency of damp earth. In the event of moulding the briquettes by machinery the quantity of water may perhaps be controlled by the amount of compacting employed and the tendency to force the water out in moulding.

The quantity of water required, as already noted, varies with the fineness, age, and other conditions of the cement, as well as with its nature.

To bring mortar to a plastic condition the quantity of water required is approximately as follows:

" For briquettes of neat cement: Portland cement about 25%, natural cement about 30%.

' For briquettes of 1 part cement, 1 part sand: about 15% of total weight of sand and cement.

" For briquettes of 1 part cement, 3 parts sand: about 12% of total weight of sand and cement."

In any particular instance the proper amount can only be determined by trial.

ART. 45. METHODS OF MAKING BRIQUETTES.

The wide differences commonly found in the results of tensile tests made by different men are, without doubt, mainly due to differences in making the briquettes. Probably if the other strength tests were as commonly used as that of tension the same want of uniformity would be observable in them. These differences occur, not only in the work of novices, but in that of skilled operators, who, while able to maintain practical uniformity in their own work, disagree in results with each other when experimenting upon the same material and apparently using the same methods. The extent of this difficulty has already been alluded to in Art. 42.

In order to secure uniform results it is essential that a uniform procedure be adopted as to all the operations of forming the briquette. The points of importance are the quantity of water used in tempering—which has been discussed in the last article, the method of gauging and amount of working to which the mortar is subjected in bringing it to a proper consistency, and the method of forming the briquette and amount of force used in placing the mortar in the mould.

In making briquettes by hand two general methods are employed, corresponding to different consistencies of the mortar. The plastic method is most commonly employed, being used in England, France, and the

United States, while the dry method is standard in Germany.

The rules recommended by the committee of the American Society of Civil Engineers give the following method of making the briquettes:

" The proportions of cement, sand, and water should be carefully determined by weight, the sand and cement mixed dry, and the water added all at once. The mixing must be rapid and thorough, and the mortar, which should be stiff and plastic, should be firmly pressed into the moulds with a trowel, without ramming, and struck off level; the mould in each instance while being charged and manipulated to be laid directly on glass, slate, or some other non-absorbing material.

" The moulding must be completed before incipient setting begins. As soon as the briquettes are hard enough to bear it, they should be taken from the moulds and be kept covered with a damp cloth until they are immersed. For the sake of uniformity, the briquettes, both of neat cement and those containing sand, should be immersed in water at the end of 24 hours, except in the case of one-day tests."

The report of the French commission upon standard tests recommends the following method:

" The moulds are placed upon a plate of marble or polished metal, which has been well cleaned and rubbed with an oiled cloth. Six moulds are filled from each gauging if the cement be slow-setting and four if it be quick-setting. Sufficient material is at once placed in each mould to more than fill it. The mortar is pressed into the mould with the fingers so as to leave no voids and the side of the mould tapped several times with the trowel to assist in disengaging the bubbles of air. The excess of mortar is then removed by sliding a knife blade over the top of the mould so as to produce no compression upon the mortar.

" The briquettes are removed from the moulds when sufficiently firm, and are allowed to remain for 24 hours upon the plate in a moist atmosphere, protected from currents of air or the direct rays of the sun, and at a nearly constant temperature of 15° to 18° C. They are then placed in the surroundings in which they are to be kept until the time for breaking. With quick-setting cements the delay is reduced from 24 hours to 1 hour for neat cement and 3 hours for sand-mortar.

" It is recommended to weigh the briquettes after removing them from the moulds to make sure of the regularity of their formation."

By the German method the mortar is mixed with less water than in the above, and the mould is filled and heaped with it. It is then rammed into place and pounded until the water flushes to the surface, after which the briquette is struck off level, and when hard enough is taken from the mould and treated as in the other case. Following are the specifications adopted by the Association of German Cement Makers:*

" On a metal or thick glass plate five sheets of blotting-paper soaked in water are laid, and on these are placed five moulds wetted with water. 250 grammes

---

* *Engineering News*, Nov. 13, 1886.

of cement and 750 grammes of standard sand are weighed and thoroughly mixed dry in a vessel; then 100 cubic centimeters of fresh water are added, and the whole mass mixed for five minutes. With the mortar so obtained the moulds are at once filled, with one filling, so high as to be rounded on top, the mortar being well pressed in. By means of an iron trowel, 5 to 8 centimeters wide, 35 centimeters long, and weighing about 250 grammes, the projecting mortar is pounded, first gently and from the sides, then harder into the moulds, until the mortar grows elastic and water flushes to the surface. A pounding of at least one minute is necessary. An additional filling and pounding in of the mortar is not admissible, since the test-pieces of the same cement should have the same density at the different testing stations. The mass is now cut off with a knife and the surface smoothed. The mould is carefully taken off, and the test-piece placed in a box lined with zinc, which is to be provided with a cover to prevent a non-uniform drying of the test-pieces at different temperatures."

For making the test-pieces of *neat cement :* " The inside of the moulds are slightly oiled, and the same are placed on a metal or glass plate without blotting-paper. 1000 grammes of cement are weighed out, 200 grammes of water added, and the whole mass thoroughly mixed for five minutes. The forms are well filled, and then proceed as for hand-work with sand-mortar.

" The mould can only be taken off after the cement has sufficiently hardened.

" The quantity of water for finely-ground or quick-setting cements must be increased."

A method somewhat in use in France for sand-mortar is that proposed by M. Candlot, and recommended by the Commission upon Methods of Testing Materials, for use with sand-mortars. It is as follows:

" Sufficient mortar is gauged at once to make six briquettes, requiring 250 grammes of cement and 750 grammes of normal sand. The weight of water necessary exceeds by 30 grammes the amount necessary to bring the cement alone to *normal consistency*."

The mortar is prepared in the ordinary manner. In forming the briquette the mould is placed upon a metal plate, and a guide fitted above it having the same section as the mould and a height of 125 millimeters.

" 180 grammes of mortar are introduced and roughly distributed in the mould and guide with a rod. By means of a metallic pestle weighing one kilogramme, and with a base of the form of the briquette but of slightly less dimensions, the mortar is pounded softly at first, then stronger and stronger until a little water escapes under the bottom of the mould.

" The pestle and guide are then removed and the mortar cut off level with the top of the mould."

It is claimed that by this method very uniform results have been obtained.

There are two points to be especially noted in making briquettes by hand: first, the mortar must be very thoroughly worked in gauging; both the German and French rules require that it shall be briskly mixed for at least five minutes, only sufficient mortar being prepared at once for five or six briquettes; second, the air-bubbles must be well worked out of the mortar in filling the moulds.   The neglect of these precautions causes much of the irregularity which commonly exists in the work of inexperienced operators.

It is perhaps easier to secure uniform results with the dry than with the plastic method.   The greater density of the hammered briquette also gives it higher strength.   The plastic method, however, accords more nearly with the conditions of the use of the material in practice.

Even with the.most experienced operators there exist differences in the amount of working, the pressure given in forming the briquette, and the quantity of water used, which cause wide variations in result.

In order to secure good results in tests of strength it is necessary that the briquettes should be kept in a moist condition during setting and the first period of hardening.   For this purpose it is customary in the United States to cover the briquettes with wet cloths after moulding and until submerging them.   In Europe they are commonly placed in zinc boxes during this period.

In the laboratory of the City of Philadelphia Mr. Richard L. Humphreys uses a soapstone closet for this purpose.   He describes * the arrangement as follows: " This closet, which is made of soapstone $1\frac{1}{2}$ inches thick, is supported on a wooden frame, and is 3 feet high and 18 inches wide.   Along the front is a strip of soapstone 3 inches wide, forming a basin of

---

the bottom of the closet, in which water is placed for keeping the air moist.   The doors are made of wood, covered with planished sheet copper, and are rabbeted to fit tightly.   There are two sets of shelves, the lower being a wooden rack, and the upper is formed of strips of glass 33 inches long, 3 inches wide. When closed the closet is perfectly tight, the water in the bottom keeping the air moist, preventing the briquettes from drying out, and thus checking the process of setting.

" Briquettes which have been removed from the moulds are placed on edge on the glass shelves, while the moulds containing the briquettes too soft to be removed are placed on the rack."

ART. 46.   MECHANICAL APPLIANCES FOR MAKING BRIQUETTES.

In order to reduce the effect of the personality of the operator in making tests of the strength of cements, various appliances for gauging and moulding briquettes by machinery have been proposed and tried.

Greater uniformity in these tests is highly desirable, and it seems possible to reach it only by the application of automatic appliances in making the briquettes.

No entirely satisfactory system of automatic testing has as yet been devised.   In Europe machines are quite commonly employed for moulding briquettes, but the mixing is done by hand in the ordinary manner.

In Germany and Switzerland all blocks for compressive tests are made by machine, as are many of the tensile specimens.   The machine used for this purpose is either the Bohmé hammer or the Tetmajer apparatus.

The *Bohmé hammer*, designed by Dr. Bohmé of the Charlottenburg Experiment Station, is shown in Fig. 11.   It consists of an arrangement by which the mor-

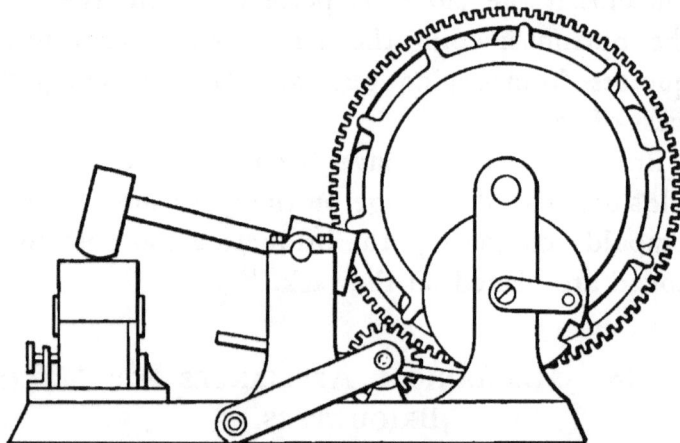

FIG. 11.

tar is compacted in the mould by a succession of blows struck by a hammer of the weight of two kilogrammes upon a plunger sliding in a guide-mold placed over the mold in which the briquette is to be formed.

The machine is arranged to lock after striking 150 blows.   A high degree of density is thus produced in the briquette, and the air is thoroughly expelled. More regular results are thus obtained, depending much less upon the personality of the operator than by the hand method.   The arrangement of this apparatus, however, is such that its operation must be extremely slow, in order to give time for the hammer to strike a full blow without being caught on the next stroke, and the time required to make a briquette is too great to admit of its general use.

The rules of the Association of German Cement Makers specify this machine, and are as follows:* " In order to obtain concordant values in compression tests machine-making is necessary. 400 grammes of neat cement and 1200 grammes of dry standard sand are thoroughly mixed dry in a vessel, 160 cubic centimeters of water are added thereto, and then the mortar is thoroughly mixed for five minutes. Of this mortar 860 grammes are placed in the cubic mould, provided with guide-mould, and the mould is then screwed on the bed-plate under the pounding-machine. The iron follower is placed in the form, and by means of Bohmé's trip-hammer one hundred and fifty blows are struck by a hammer weighing 2 kilogrammes. After removing the guide-mould and follower the test-piece is smoothed off, taken with the mould from the bed-plate, and then treated as for hand-work."

The *Tetmajer apparatus* is similar in character to the Bohmé hammer. It consists of an iron rod carrying a weight upon its lower end, which is raised through a given height and dropped upon the mortar in the mould. The ram in this machine weighs 3 kilograms. This machine is used in the Zurich laboratory for both tensile and compressive specimens, and Prof. Tetmajer regulates the number of blows by requiring a certain amount of work to be done upon a unit volume of mortar,—0.3 kilogrammeter of work per gramme of dry material of which the mortar is composed.   This apparatus is subject to the same limitations in practice as the Bohmé hammer, in being very slow in use, and somewhat expensive in first cost of apparatus.

Canadian Method.—In Canada, and to some extent in England, the method has been adopted of gauging the mortar quite soft, using a high percentage of water, with machine-mixing, and then moulding the briquettes under light pressure—20 lbs. per square inch on the surface of the briquette.   This gives much lower results for strength than the ordinary methods, which is unimportant provided the results are concordant, and specifications are made to agree with the method.

A number of tests made by Mr. Cecil B. Smith at the McGill University seem to show that the method is capable of yielding uniform results in so far as the variations of the individual tests from the mean of a single series is concerned.   The method appears to be defective, however, in not affording a satisfactory method of determining the proper consistency of the mortar, upon which largely depends the comparability of the results of different observers.

The Jamieson briquette machine, shown in Fig. 12, was designed by Prof. Jamieson, of the Iowa State University, for the purpose of making briquettes under diect pressure, and is intended to secure rapid manipulation.

The following description is from *Engineering News* of February 7, 1891: " The principle of the machine is very simple.   A vertical cylinder (*a*) whose cross-

section conforms to the outline of the standard briquette receives a charge of the mixed cement from a

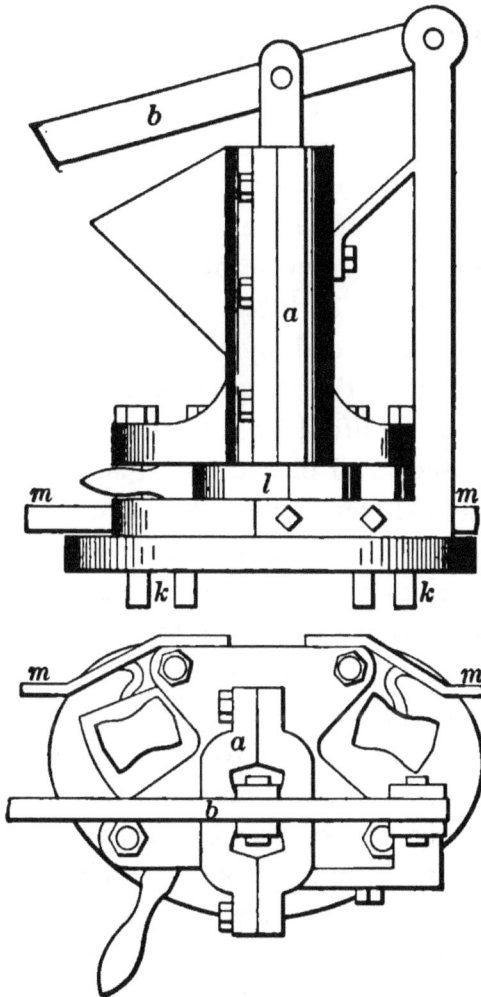

FIG. 12.

hopper at the side. In this cylinder a close-fitting
piston is worked by a hand-lever (*b*), so that the

charge of cement in the cylinder may be subjected to
, a pressure of about 175 lbs. per square inch. The
bottom of the cylinder is raised above the bed-plate a
distance equal to the thickness of the briquette (1 inch),
and in this space works a triangular block (*l*) having
two holes of the same cross-section as the briquette.
This block can be oscillated to bring either of these
holes beneath the cylinder. When one hole is in this
position, however, the other is clear of the cylinder at
one side, and the briquette of cement which has been
pressed into it can be lifted by a plunger, worked by
·a lever (*m*) and guided by the pins (*kk*). Owing to
the pressure used the briquettes are hard enough to
handle as soon as lifted from the mould, and they are
at once removed and placed on a glass slab.''

" In practical working it has been found possible
to make briquettes as rapidly as 600 per hour.''

This machine may be found useful in laboratories
where large numbers of briquettes are needed for the
purpose of comparing cements under various conditions, as it greatly lessens the manual labor of forming the briquette. It is to be observed, however,
that the time occupied in making briquettes is largely
used in mixing the mortar, and that rapid moulding
involves equally rapid mixing. The macnine, as
designed, provides no means of regulating the amount
of pressure applied in forming the briquette, which
may vary with different operators, and is thus likely
to produce variations in result. This may be comparatively unimportant for neat briquettes mixed dry,
but has a large influence upon the strength of sand-
mortar. It may be suggested also that to make

by this method briquettes, which are firm enough to handle immediately, the cement must be mixed very dry—too dry for the best results; with sand-mortars it will be very difficult to produce solid cakes by this method.

The experience of the author in experimenting with various appliances for moulding briquettes shows that quite uniform results may be obtained from briquettes moulded under a single application of a steady pressure.

This method may be applied much more rapidly than the hammer method, with about as good results, and render the results of different operators much more concordant than can be obtained by hand-work, although the variations from the mean in the work of a single observer may be as great or greater than in the hand-work.

Briquettes of neat cement, machine-mixed, and moulded under a pressure of about 500 lbs., upon the surface of the briquette, give good results when the averages of different men are compared, and small variations in pressure are not important in the results.

For sand-mortar, one part cement to three parts sand, a pressure of 1000 to 1500 lbs. is desirable to sufficiently compact the mortar to form homogeneous briquettes and give uniformity in the results.

To obtain the best results the mortar should be gauged to such a consistency that the water begins to ooze out under the pressure, the cakes being reduced to a semi-plastic condition, and becoming too soft to handle before setting.

An apparatus for making briquettes by this method

is easily arranged. To do good work it should not aim at too great rapidity, but sufficient time should be given to each briquette to permit the weight to produce its full effect in the compression of the mortar. The gradual application of the load in a testing-machine has been found preferable to the sudden descent of a lever as usually employed. Possibly a screw and hand-wheel giving a gradually increasing pressure may prove the best method of applying the pressure. The time occupied need not be long, and briquettes may be made much more rapidly than in hand-work. A means of regulating the pressure so that it may always be the same is an essential to good work.

It is quite as important to eliminate the personal element from the gauging as from the moulding of briquettes. For this purpose several appliances have been tried.

The *Jig mixer* is an apparatus in which the materials are placed in cups with covers clamped on, and shaken rapidly up and down. It has been tried in a number of places, but has usually been found quite unsatisfactory in practice. It is difficult to make a satisfactory mixture by this method, and the result depends very much upon the rapidity of operation.

The *Faija mixer* was designed and first used by Mr. Faija in England. It is shown in Fig. 13, as made by Riehle Bros. Testing Machine Company of Philadelphia, and consists of a cylindrical pan in which a mixer, formed of four curved blades, revolves both on its own axis and about that of the pan. This arrangement gives fairly good results in use.

Fig. 14 shows an arrangement used by the author

in the laboratory of the College of Civil Engineering at Cornell University.

In this apparatus the cylinder containing the materials is closed, thus avoiding the dust, which is very disagreeable with the open pan. The cover revolves about the axis of the cylinder, and the gear is placed

 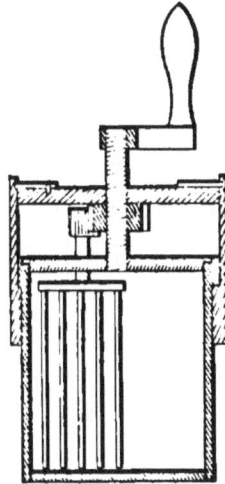

FIG. 13. FIG. 14.

outside to remove any danger of it becoming clogged. The mixer is formed of vertical rods held by a horizontal arm, and revolves about the axis of the cylinder and also about the middle point of the arm. This arrangement seems to give a somewhat more thorough working of the mortar than the same number of turns of the Faija mixer, but if the briquettes be moulded under heavy pressure the two give about the same results.

By the use of some apparatus of this kind the mortar may be much more expeditiously mixed than by hand, and with greater uniformity.

To devise any system of making briquettes by hand which will secure uniformity in the work of different men seems hopeless.   If such uniformity is to be secured, it must be by the use of automatic appliances. This involves not only the use of the same apparatus, but its use in the same manner in the different labora-tories.   This would be very difficult of attainment; but if various mechanical appliances should come into use the relations between them would soon become known, and statements of the means employed in form-ing briquettes would make the results to a certain extent, at least, comparable with each other.

For the purpose of securing uniformity in the results of pressure-made briquettes the following pro-cedure has been followed, with good results, in the laboratory of the College of Civil Engineering at Cor-nell University: In gauging the mortar two pounds of the dry materials are employed in one mixing.   For sand-mortar, the dry materials are put in the mixer and given 50 turns; the water is then added and the mixing completed by an additional 50 turns, the mixer being operated at the rate of about 100 turns per minute.   With neat cement, the cement is put in the mixer, the water added, and the mixing completed with 50 turns of the handle.   After the completion of the mixing the mould and guide are filled with the mortar, and a pressure of 1000 lbs. put upon the piston in the pressure-machine.   The piston is then raised, the guide removed, and the surplus of mortar sliced off; after which the briquette is removed from the mould by being pressed out upon the surface upon which it is to remain until set.   The necessary quantity of water is first determined by trial, and is such that the pressure used will cause the water to slightly ooze out beneath the mould.

ART. 47. TENSILE TESTS.

The test for tensile strength is commonly made by placing the briquette in a pair of clips which catch its ends and are attached to a machine by which the load necessary to break the briquette may be weighed. In order to secure uniform results it is necessary that the stress shall be so applied as to bring the tension axially upon the small section of the briquette, and also that the rate of application of the load shall be always the same.

There are various types of testing-machines in use, and it seems unnecessary to enter into any detailed description of them here. One of the simplest is the old machine in the College of Civil Engineering at Cornell University, which consists of a lever, suspended from a wooden framework, carrying near one end the clips to hold the specimen, and at the other a bucket into which a stream of water flows until the briquette breaks, when the water is shut off automatically. The weight of water multiplied by the ratio of lever-arms gives the stress on the briquette. All of the points of suspension are on knife-edges, and an adjustable weight upon one end serves to balance the lever.

The *Michaelis Machine*, shown in Fig. 15, is similar in character to the above, but is arranged with a double lever, and uses shot in the bucket instead of water. This machine is the one commonly employed in Europe.

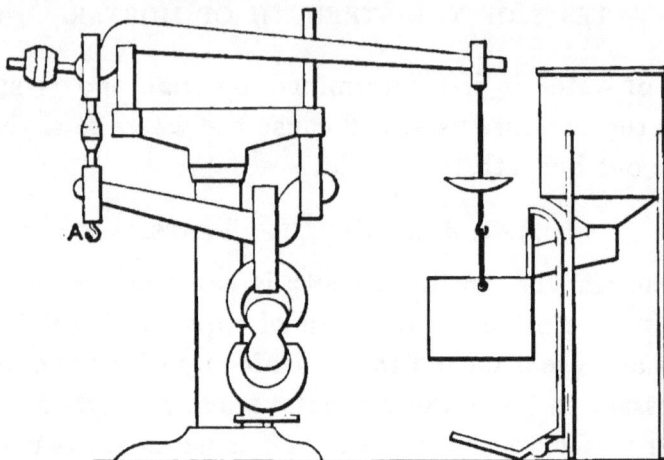

FIG. 15.

FIG. 16.

The *Fairbanks Machine* is practically the same as that of Michaelis. In the Fairbanks machine the shot is weighed and the stress determined by the same beam used in breaking the specimen, the bucket being hung at the other end of the beam from that used in breaking the specimen, and the weight found by means of a sliding weight.

The *Richlé Machine*, shown in Fig. 16, is an ordinary lever machine in which the load is brought upon the specimen by means of the lower hand-wheel, while the weight is moved along the scale-beam by the upper hand-wheel. In testing a briquette both wheels must be operated simultaneously and the scale-beam be kept balanced.

The *Olsen Machine*, shown in Fig. 17, is similar in character to the above, but differs somewhat in detail. This machine has been modified by Prof. Porter in one constructed for Lafayette College,* by adding a second lever below the clips, through which the stress is applied by means of water flowing into a bucket attached to its end. The stress is measured by the weight sliding upon the upper scale-beam as in the ordinary machine, but the weight is moved automatically by means of an electric contact at the end of the beam.

Various other more complicated types of machines are sometimes employed, most of them hydraulic in principle.

Nearly any of the machines in common use may give good results in practice. In selecting a machine,

---

* *Engineering News*, March 5, 1896.

however, those are to be preferred in which the load
may be automatically applied at a uniform rate.   The

FIG. 17.

attainment of a constant rate of application with a
hand-machine is a matter of considerable difficulty.

In order that the stress upon the briquette shall be
axial, care must be exercised in properly centring the

briquette in the clips, and the form of the clip must
be such that it shall not clamp or bind upon the head
of the briquette, but may be free to adjust itself to an
even bearing.   The surface of contact between the clip
and briquette must be large enough to prevent the
material of the briquette being crushed at the point of
contact, and yet as small as possible to admit of its
more free self-adjustment.   The suspension of the
clips, as is usual, by conical bearings permits them to
turn so as to always transmit the stress in a right line
between bearings.

Fig. 18 shows the form of clip which is used with

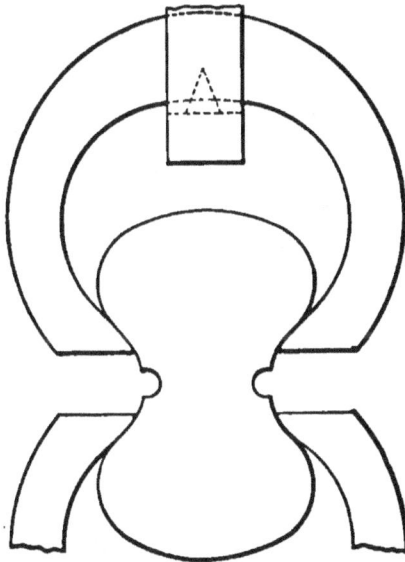

FIG. 18.

the standard German briquette.   As here given it is
approved by the " Commission des Méthodes d'Essai
des Matériaux de Construction," as giving very satisfactory results.

With the briquette used in the United States and

England several forms have been used for clips.   Fig.
19 shows the form adopted by the Committee of the
American Society of Civil Engineers.   This form does
not offer sufficient bearing-surface for good results, as
the briquette is likely to break on account of the
crushing of the surface of the briquette at the point
of contact.

The form shown in Fig. 20 has been much used, but

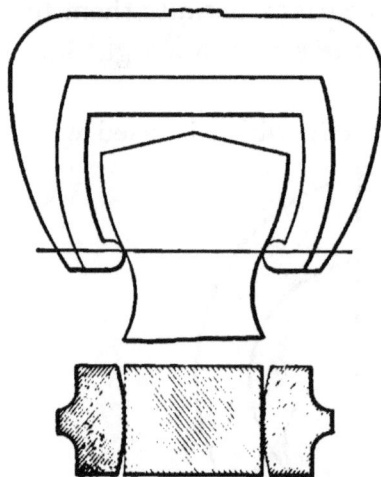

FIG. 19.   FIG. 20.

has the disadvantage of clamping the head of the
briquette too closely, and unless great care is used
may cause the briquette to break by twisting, thus
giving irregular results. · When the break occurs ·
through the crushing of the material the fracture
usually extends from one of the points of contact
irregularly to the small section, but a break due to
twisting may often be a centre break, and the irregularity show only in the results.

The form of clip shown in Fig. 21 is to be preferred
to those given above, as affording a sufficient bearing-

surface without clamping, and thus tending to diminish the irregularities in the results of tests.

In order to prevent crushing at the points of contact

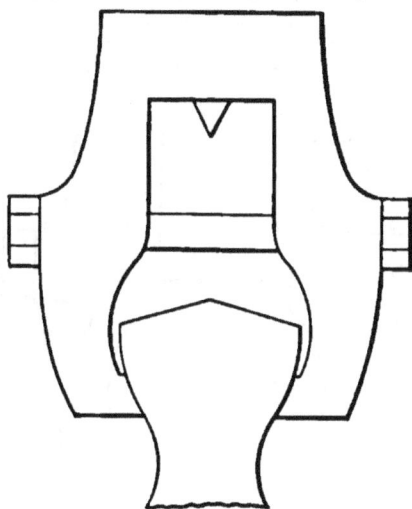

FIG. 21.

and to permit more free adjustment Mr. W. R. Cock has proposed * a rubber bearing, as shown in Fig 22. The use of this clip undoubtedly increases to some extent the proportion of centre breaks, and perhaps slightly raises the breaking strength. A rubber bearing-surface may also be secured on the clips of form shown in Fig. 21, by stretching a rubber band around the jaw of the clip so as to cover the surface of contact. This may somewhat increase the regularity of the tests, but when carefully used the ordinary clip is capable of giving very satisfactory results.

Various appliances have been proposed for the accurate centring of briquettes, or to prevent the more free adjustment of the clips to the direction of stress, by

* *Engineering News*, Dec. 20, 1890.

using a template for the exact placing of the briquette,
or by hinging the clips at the upper corners.   These
arrangements do not, however, seem to be necessary.

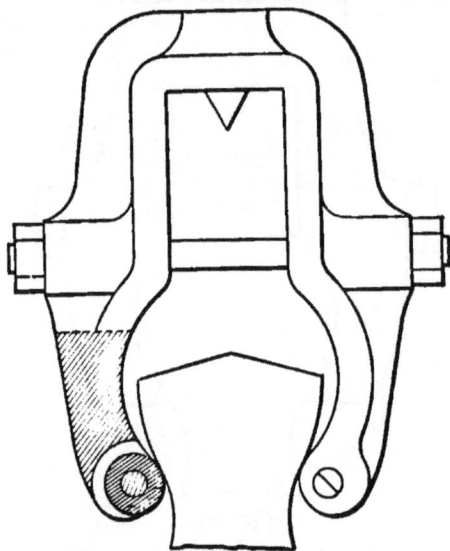

FIG. 22.

The method of keeping the briquettes between the
time of moulding and breaking is of course important
in its effect upon the resulting strength.   The temperature should, for standard tests, always be kept as
nearly uniform as possible—between 60° and 70° Fahr.
When the mortar has been conserved under water it
should be tested immediately on being taken from the
water.

The standard rate of applying the load is ordinarily
about 400 lbs. per minute.

ART. 48.   COMPRESSIVE TESTS.

The compressive strength of cement-mortar is much
greater than its tensile strength, and as it does not

seem to give a better indication of value, while more difficult of satisfactory determination, and also requires heavier apparatus, it is not usually employed as a test of quality in the acceptance of material.   The compressive test is, however, valuable for purposes of comparison, and is desirable as an addition to the showing made by the tensile test.   In the European experiment stations it is customary to test all cements under compression as well as tension.

For this test, as for tension, it is essential that a standard method be followed if comparable results are to be obtained.   The size and shape of the specimen are of the greatest importance.   The effect of compression is ordinarily to cause the material to spread laterally by pressing out the sides, failure usually occurring by shearing along surfaces inclined at about 30° with the vertical, leaving pyramidal or conical blocks at the middle.

When the specimen is small in height the resistance is greater per unit area than if it be higher, and for blocks similar in form the resistance increases with the size.   Cubes are commonly employed for this purpose, but cylinders are sometimes preferred, on account of the greater ease with which homogeneous specimens may be prepared, and because of the liability of the corners of cubes to crack off under comparatively light pressure.

The common piece in Europe is a cube with edges 7 centimeters long, moulded under the hammer, and in testing it is required that the pressure be always exerted on two faces of the cube, that is, on the faces which are against the surface of the mould, in forming the block.

The German conference at Berlin, however, recommended the use of blocks 5 square centimeters in area, the same as the tensile specimens.

The French " Commission on Methods of Testing Materials " made the following recommendations:

" For tests of compressive strength the half-briquettes separated by tension are to be used. Each half-briquette to be crushed separately, and the sum of the two results taken as the strength of the specimen.

" In the absence of half-briquettes cylinders 45 millimeters in diameter and 22 millimeters high may be used, made and conserved like the tensile briquettes.

" Those specimens which show visible irregularities or distortions are smoothed by lightly rubbing in the hand upon a stone surface."

" The testing apparatus should be so arranged that the stress may be continuously applied at such a rate as to crush the half-briquette in one or two minutes."

" For tests having for their object the comparison of mortar with other materials, it is provisionally recommended to employ the cube, with faces of 50 square centimeters area, placed upon one side. These tests will thus conform in a general way to the rules adopted for the other materials."

In testing cement the compressive specimens may be readily obtained with smooth faces, which may be placed in the jaws of the testing-machine directly in contact with the compression-blocks. This is the usual practice in Europe.

It is common, however, in testing other materials, stone or brick, to set the specimen in the machine

with a thin layer of plaster of Paris between the plate
of the machine and the surface of the specimen.   A
small pressure being placed upon the specimen before
the plaster of Paris sets, the bearing becomes even and
the load uniformly distributed over the surface.   In
many instances this seems to conduce to greater
uniformity in results with cement, but it is not common practice and is hardly necessary.

For making compressive tests of cement, any of the
ordinary lever or hydraulic machines in common use,
with a capacity of 40,000 to 50,000 lbs., is usually
sufficient.   It is desirable that the load be applied as
uniformly as possible, as the result will be more or
less affected by unsteadiness or shocks.

A very handy and efficient machine for this purpose
in use at the College of Civil Engineering, Cornell
University, is shown in Fig. 23.   It is a hydraulic
machine, constructed by J. Amsler, Lafon & Son,
Shifflausen, Switzerland.   The machine is a hydraulic
press, in which fluid pressure is reduced through a
system of pistons, so that it can be measured by a
mercury manometer. A is the compression-piston, B
and C are the pistons which reduce the pressure existing under piston A.   The mercury manometer consists
of a glass tube connected at the bottom with the space
under the piston C.

The test-piece E is placed between the compression-
plates, of which one rests upon a spherical surface in
the piston A, so that it can adjust itself; the other
hangs on the end of the hand-screw, and may be placed
at any height.

The cylinder K is filled with castor-oil.   The rod

L may be driven into the cylinder by the crank and
screw, producing a pressure on the oil between pistons
A and B. The piston A presses upward on the speci-

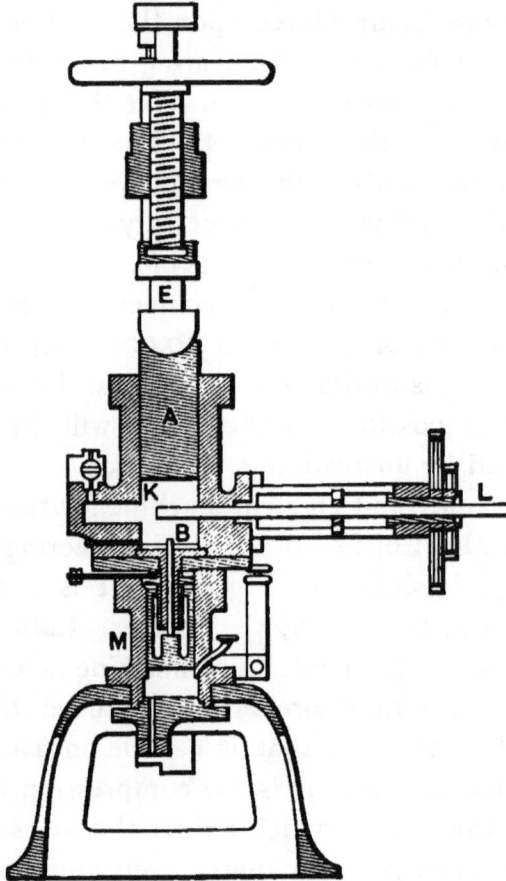

FIG. 23.

men, while piston B presses downward upon the larger
piston C, causing pressure on the fluid beneath.
Under piston C is a layer of machine-oil, which serves
the purpose of making it tight and lubricating it.

The mercury fills the bottom of cylinder M (under

the machine-oil) and the pipes connecting this cylinder with the manometer, which stands at the side of the machine and reads to 32,000 kilogrammes.

The pistons are fitted accurately to the cylinders, and are not packed. When the machine is operated by turning the crank which drives the rod L the pistons C and B are caused to oscillate about their axes, thus equalizing the wear and eliminating the friction. This machine is easily operated by hand, and the speed may be regulated by watching the manometer while turning the crank.

It has been proposed to employ punching tests instead of crushing the entire specimen. This method has been employed for a number of years in the laboratory at Teil, France. A punch five square centimeters in section (circular) is employed, and it is claimed that the results are more regular than those obtained by crushing the entire specimen, while requiring less force in the testing-machine. The " Commission des Méthodes d'Essai des Matériaux," however, concluded that it presented no advantage over the ordinary test.

ART. 49.   TRANSVERSE TESTS.

Tests of the strength of mortar under transverse loading are seldom employed as a measure of the quality of the material, but are frequently made with a view to determining the action of the material in service.   Propositions have often been made to substitute the transverse for the tensile test in the reception of material.   These suggestions have usually been based upon the simplicity of the test and of the apparatus with which it may be carried out. All that is necessary, after the bar is prepared, is the arrangement of a couple of knife-edges upon which the ends of the bar may be rested, and a third knife-edge to carry the weight brought upon the middle of the bar. The ordinary test by tension is, however, quite simple, and there seems to be little if any advantage in making a change, although the transverse test offers an equally effective means of determining quality. Much less is known as to what the transverse strength should be, and its use in specifications would need to be preceded by experiments to obtain proper values for the loads to be required, while the errors due to differences in making briquettes would exist the same in the one case as in the other.

Prof. Durand-Claye made a large number of tests to compare tensile and transverse strength for both neat cement and mortars in small test-pieces. He used bars 2 centimeters square and 12 centimeters long, and tension-pieces of the ordinary 5-square-centimeter section, and found the results quite regular. The modulus of rupture per square centimeter, computed by the ordinary formula $(R = \dfrac{3}{2} \dfrac{Pl}{a^3}$, where $P =$ load, $l =$ length, and $a^2 =$ area), were found to average a little less than twice the unit stress for tension. The use of this formula in this instance is of course inexact, as it assumes the material to have the same coefficient of elasticity for tension as for compression, and to be strained only to the elastic limit. As all the specimens are of the same size, however, this is immaterial for purposes of comparison.

In making the transverse test the most common method has been to provide supports for the ends of the bar and hang weights directly upon its middle. Care should be exercised to guard against the crushing of the material under the knife-edges; a good plan is to use small plates of iron between the knife-edge and the surface of the briquette to distribute the load.

Where a tension-machine is in use, a transverse attachment is readily added, by which both tests may be made by the same machine. Fig. 24 shows such

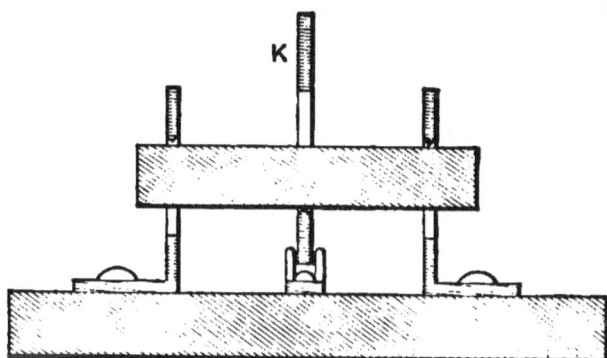

FIG. 24.

an attachment as it is commonly used in Europe in connection with the Michaelis machine; the rod K is attached to the end of the lower lever in the machine (A, Fig. 15), thus receiving a less effect from the application of the load than the tension specimen.

ART. 50.   TESTS OF SAND-MORTAR.

Tests of the strength of sand-mortar, although commonly recommended, are not very generally employed in ordinary specifications, reliance being usually placed upon the neat test, coupled with that for fineness, to indicate how the cement will act when mixed with sand. A tensile test with sand is, however, of the greatest value when properly conducted, because it accords more nearly with the conditions under which the cement is to be used.

Tests with sand may be made either as a means of judging the value of the cement, or to determine the probable strength of mortar under the circumstances which may obtain in special work. When the tests are intended as a determination of value, it is essential that they be made according to some standard forming ready means of comparison with other material. This requires that the specimens be made and tested according to standard methods, and that the sand used be of standard quality.

In the German specifications the standard sand is described as follows: " In order to obtain concordant results in the tests, sand of uniform size of grain and uniform quality must be used. This standard sand is obtained by washing and drying the purest quartz sand obtainable, sifting the same through a sieve of 60 meshes per square centimeter, thereby separating the coarsest particles, and by removing from the sand so obtained, by means of a sieve of 120 meshes per square centimeter, the finest particles. The diameters for the wires of the sieve shall be 0.38 millimeter and 0.32 millimeter respectively." The German Conference upon Uniform Tests specified Freienwalde sand as the standard.

The committee of the American Society of Civil

Engineers recommend an artificial sand made by crushing quartz.   Their report is as follows:

" The question of a standard sand is of great importance, for it has been found that sand looking alike and sifted through the same sieve gives results varying within rather wide limits.

" The material that seems likely to give the best results is the crushed quartz used in the manufacture of sand-paper.   It is a commercial product, made in large quantities and of standard grades, and can be furnished of a fairly uniform quality.   It is clean and sharp, and although the present price is somewhat excessive (3 cents per pound), it is believed that it can be furnished in quantity for about $5.00 for 300 lbs. As it would be used for tests only, for purposes of comparison with the local sands and with tests of different cements, not much of it would be required.

" The price of German standard sand is about $1.25 for 112 lbs. ; but the article being washed, river-sand is probably inferior to crushed quartz.   Crushed granite could be furnished at a somewhat less rate than crushed quartz, and crushed trap for about the same as granite; but no satisfactory estimate has been obtained for either of these.

" The use of crushed quartz is recommended by your committee, the degree of fineness to be such that it will all pass a No. 20 sieve and be caught on a No. 30 sieve."

This sand can now be obtained of dealers in testing apparatus, but much of what is furnished for the purpose requires resifting to bring it to standard size. The use of this sand is probably advantageous as conducing to uniformity, but it gives less strength in mortar than good natural sand.

In France both artificial sand (crushed quartz) and natural sand are used to some extent in tests. The Commission on Methods of Testing recommend the use of natural sand. They divide the sand into these sizes:

1. Sand which passes openings of 1 mm. and is retained by those of 0.5 mm.   2. Sand which passes openings of 1.5 mm. and is retained by those of 1.0 mm.   3. Sand which passes openings of 2.0 mm. and is retained by those of 1.5 mm.

The name *simple standard sand* is given to No. 2, and the name *compound standard sand* to a mixture of equal parts of the three sizes. The former is to be used in making mortar for standard tests where the dry consistency is employed, and the briquettes made by pounding. The latter is required in gauging when the plastic consistency is used.

Tests of sand-mortar for the purpose of comparing various sands with the standard sand, or of estimating the efficiency of mortar under varying circumstances of use, are often of much value as a guide to the proper use of the material. In such work the method of testing necessarily depends upon the point to be investigated.

ART. 51.   INTERPRETATION OF RESULTS.

The test for strength is regarded as a measure of the value of a cement, as showing the possession of the active elements. There are, however, different elements which act at different rates, and it is unwise to classify cements according to strength alone. The tensile strength developed by cement in a test extending over a short period of time is not necessarily an indication of the strength that may be attained by it during a longer period, unless the normal action of the particular material be known. That which is strongest at first may not continue to be the strongest.

The development of good strength soon after the use of the mortar is a desirable attribute in most engineering work, and the probability of the material being good is greater where it shows a fair early strength, and therefore it is usually wise to specify that the cement shall develop a fairly good strength on a short-time test, but there is no object in requiring extremely high values.

Mr. Faija recommends that the gain in strength between the 7 and 28 day periods be considered, rather than the absolute early strength, in determining the probable subsequent gain in strength. This is doubtless a better guide than the usual one, but it is not usually practicable to require tests extending over a period of 28 days, and in many instances it would be misleading, if comparison of different cements were attempted.

Prof. Unwin gives a formula for the strength at any period, $y = a + b(x - 1)^n$, in which y is the strength at x weeks after mixing; a, the strength at end of one week; n, a constant for the particular material, to be determined by observations extending over considerable time; b, a constant to be determined from the strengths given by the sample at 1 week and 4 weeks after mixing.   Prof. Unwin gives the value $n = 1/3$ for Portland cement in general, and shows that the formula gives values well in accord with the results of tests in many instances.   The formula depends upon the assumption that for any two Portland cements the gains in strength at end of any period are to each other as the gains between the 7 and 28 day test,—a proposition quite wide of the mark in many instances.

The strengths commonly required by specifications, in the United States, are based upon the recommendations of the committee of the American Society of Civil Engineers, which gives the following average values:

' American Natural Cement, Neat.

1 day, 1 hour or until set in air, the rest of the 24 hours in water, from 40 to 80 lbs.

1 week, 1 day in air, 6 days in water, from 60 to 100 lbs.

1 month, 1 day in air, 27 days in water, from 100 to 150 lbs. ·

1 year, 1 day in air, remainder in water, from 300 to 400 lbs.

American and Foreign Portland Cements, Neat.

day, 1 hour or until set in air, the rest of the 24 hours in water, from 100 to 140 lbs.

1 week, 1 day in air, 6 days in water, from 250 to 550 lbs.

1 month, 1 day in air, 27 days in water, from 350 to 700 lbs.

1 year, 1 day in air, remainder in water, from 450 to 800 lbs.

American Natural Cements, 1 *part Cement to* 1 *part of Sand.*

1 week, 1 day in air, 6 days in water, from 30 to 50 lbs.

1 month, 1 day in air, 27 days in water, from 50 to 80 lbs.

1 year, 1 day in air, remainder in water, from 200 to 300 lbs.

American and Foreign Portland Cements, 1 *part of Cement to* 3 *parts of Sand.*

1 week, 1 day in air, 6 days in water, from 80 to 125 lbs.

1 month, 1 day in air, 27 days in water, from 100 to 250 lbs.

1 year, 1 day in air, remainder in water, from 200 to 350 lbs."

At least the minimum value here given for 1 and 7 days are usually required in ordinary specifications, but higher values are often employed. Many of the better cements commonly give results above the maximum values stated above; this depends, however, upon how they are tested.

The values to be required in specifications need to be modified, especially with the natural cements, to accord with the particular kind of cement to be used, and also with the practice of the laboratory.

Where large quantities of cement are regularly employed, and the same men continuously make the tests, it is a comparatively simple matter to conform the specifications to the work of the laboratory so as to get reliable indications of the value of the material. A very large portion of the testing for reception of material must, however, be done upon detached works, where temporary laboratories are to be used, and inspectors employed for the occasion. In these cases it is a difficult matter to adopt a specification which shall give good results, unless the operator can himself first be calibrated.

The results of tests in the permanent laboratories usually give higher strength for the same material than would be obtained on an ordinary outside test, especially by a comparatively inexperienced man. It is not to be inferred, however, that the highest results are necessarily the outcome of the greatest skill. As a rule, the most expert and reliable operators get only moderate strength for the best material.

Lack of skill in conducting tests nearly always tells against the material tested, and good material may often be rejected because of inexperience in the inspector; but, on the other hand, it is a frequent trick of contractors having inferior material rejected on an ordinary test to send it to one of the laboratories known to obtain abnormally high strengths, and thus get results which seem to show error on the original test.

In specifications it is usually desirable to require tests showing a fair degree of strength rather than very high values. The latter are if anything less likely to give good material and unnecessarily limit competition.

CHAPTER VII.

TESTS FOR SOUNDNESS.

ART. 52 ORDINARY TESTS.

SOUNDNESS is the most important quality of a cement, as it means the power of the cement to resist the disintegrating influences of the atmosphere or water in which it may be placed. Unsoundness in cement may vary greatly in degree, and show itself quite differently in different material. Cement in which the unsoundness is very pronounced is apt to become distorted and cracked after a few days, when small cakes are placed in water. Those in which the disintegrating action is slower may not show any visible change of form, but after weeks or months gradually lose coherence, and soften until entirely disintegrated.

The method in common use for testing unsoundness is to make small cakes or pats of neat cement, usually about 3 or 4 inches in diameter and 1/2 inch thick, upon a plate of glass, and keep them in air or water for a few days, carefully watching them to see if they show any signs of distortion or surface cracks, which may indicate a tendency to disintegration.

The German standard specifications require that the

cakes for this test shall be 1.5 centimeters thick at the centre and have thin edges.   These cakes are placed in water 24 hours after they are made, or at least not until they are firmly set, and observations are continued over a period of 28 days, when, if no cracks or distortions appear, the cement is considered sound.

" The cakes, especially those of slow-setting cement, must be protected against draughts and sunshine until their final setting.   This is best accomplished by keeping them in a covered box lined with zinc or under wet cloths.   In this manner the formation of heat cracks is avoided, which are generally formed in the centre of the cake, and may be taken by an inexperienced person for cracks formed by blowing."

The Committee of the American Society of Civil Engineers recommend the following:

"Make two cakes of neat cement two or three inches in diameter, about one half inch thick, with thin edges. Note the time in minutes that these cakes, when mixed with water to the consistency of a stiff plastic mortar, take to set hard enough to stand the wire test recommended by General Gillmore, 1/12-inch diameter wire loaded with one fourth of a pound, and 1/24-inch loaded with one pound.

" One of these cakes, when hard enough, should be put in water and examined from day to day to see if it becomes contorted, or if cracks show themselves at the edges, such contortions or cracks indicating that the cement is unfit for use at that time.   In some cases the tendency to crack, if due to free lime, will disappear with age.   The remaining cake should be kept in the air and its color observed, which for a good cement should be uniform throughout, yellowish blotches indicating a poor quality the Portland cements being of a bluish gray and the natural cements being dark or light, according to the character of the rock of which they are made. The color of the cements when left in the air indicates the quality much better than when they are put in water.''

The color test above given is not of much value, as unsound cement is very commonly of good color. The time during which these observations shall continue is not specified in these rules, but in practice they are not usually carried over more than two days to a week before acceptance of the material.

The French Commission upon Methods of Testing Materials recommend both a pat test and test of the amount of swelling which takes place in the mortar, as follows:

'' *Cold Tests.*—(*a*) For this test the cement paste is formed into a pat about 10 centimeters in diameter and 2 centimeters thick, made thin at the edges.

'' Immediately after being made the specimens intended for tests in water are immersed in the same conditions as the briquettes used for tensile tests.

'' The specimens intended for use in the air are at once exposed to the conditions indicated for briquettes.''

'' The condition of the specimens is observed at the same periods of time employed in making tensile tests (7 days, 28 days, 3 months, 6 months, 1 year, etc.).

'' (*b*) To measure the increase in volume of the mortar of neat cement after prolonged immersion in cold water, a bar of cement is employed 80 centi-

meters in length and 12 millimeters square, placed
vertically in a glass tube 25 millimeters in diameter,
which is then filled with water.''

'' The elongation is measured by the motion upon
an arc of a needle moved by a rod resting upon the
upper extremity of the bar of mortar.'' (See Fig. 25.)

The method of testing by measuring the variation
in length is also used to some extent in Germany.
Methods of conducting this test are described in Art.
53.

It is important in testing soundness in this manner
that the tests should be continued over as long a
period as possible, and many cases of unsoundness are
not discovered with a 28-day test.   Instances have
been observed in which mortar in the form of 2-inch
cubes has completely disintegrated within two years,
where incipient checking was not observable for three
months in a small cake test.   The most common and
dangerous cases of unsoundness are probably discovered by the ordinary tests.   It may be observed, however, that the fact that disintegration of mortar is not
oftener observed in large constructions is probably due
more to the general good quality of the cement supplied by the best makers, and to the frequent stability
of work regardless of the nature of the mortar, than
to the efficiency of the test for soundness.

The quantity of water to be used in mixing mortar
for tests of soundness is about the same as that used
for tests of strength, although a variation in the quantity, within small limits, does not seem to materially
influence the results.   Care must be taken that the
cakes be kept moist during setting and previous to

immersion, in order that they may be free from drying cracks. On this account the French commission consider it preferable to immerse the specimens immediately after mixing, without waiting for the mortar to set. Some natural cements do not stand immediate immersion although apparently quite sound in water if first allowed to set in air.

When mortar is to be used in sea-water, the pats of cement are placed in water of the same character, and as nearly as possible corresponding to the conditions of practice.

Practically, however, the action of sea-water is so slow that the test is comparatively useless. M. Alexandre found * that the first indication of disintegration may not be shown for several months or perhaps years. He also found that the action in the laboratory did not always accord with that in the work. This was probably due to failures occurring because of the method of employing the mortar, when the cement was not defective.

Tests are sometimes made of mortar to be used in sea-water by causing the water to filter through the block of mortar under pressure. This test is made in France by employing the standard cubes used in compression, which are arranged and submitted to filtration as in the tests for permeability (see Art. 67). These blocks are afterward crushed, and their strengths compared with those kept under normal conditions.

---

* Annales des Ponts et Chaussées, Sept. 1890, p. 131.

ART. 53.   MEASUREMENT OF EXPANSION.

Unsoundness in cement is doubtless for the most part due to the presence of expansive elements, the action of which subsequent to the setting of the cement produces internal forces, tending to disrupt the mass of mortar and usually causing an increase in its volume.   The various tests for soundness have for their object the determination of the presence of these expansives, which may be indicated either by the distortion and cracking of the mortar when present in sufficient quantity, or by a simple increase in volume without visible distortion when present in less quantity, or when more finely divided and uniformly distributed through the mass.

As a more efficient and accurate method of determining the presence of expansives than is afforded by the observation of thin pats of mortar, various appliances have been devised for the purpose of measuring directly the increase of volume of a block of mortar.

The Long-bar Apparatus.—The first apparatus for the purpose of measuring change of dimensions was devised by MM. Durand-Claye and Debray.   It has been in use for a number of years in France, and was recommended for use in cold tests by the French Commission on Methods of Testing Materials (see Art. 52).   This apparatus is shown in Fig. 25.   The test is made upon bars 80 centimeters long and 12 millimeters square in section.   The moulds in which these bars are formed consist of iron rods considerably longer than the bars to be formed, and of section 30 by 12 millimeters, laid flat upon a table, held apart at the ends by blocks of the same section as the cement bar, and prevented from spreading by clamps at the ends.

In making the test, the bar of cement is placed in a vertical glass tube 80 centimeters long and 23 mil

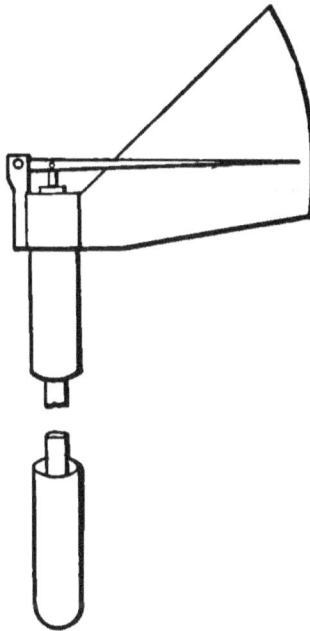

FIG. 25.

limeters in diameter, closed at the bottom and filled with the water. to the action of which the mortar is to be exposed.   To the top of the glass tube is fixed a ring carrying on one side an arc and on the other a rod, to which is hinged a needle, which travels upon the arc, being actuated by a rod resting upon the top of the cement bar.   The extension of the cement bar is thus multiplied by 10 in the reading on the arc, which may be graduated, and the successive positions

of the needle read from the scale, or it may be covered
with blank paper, and have the positions of the needle-
point marked and afterward measured.   This method
requires much care in manipulation, both in making
the bars and in handling them in the test.

Bauschinger's Caliper Apparatus.—This apparatus,
designed by Prof. Bauschinger and used in the German

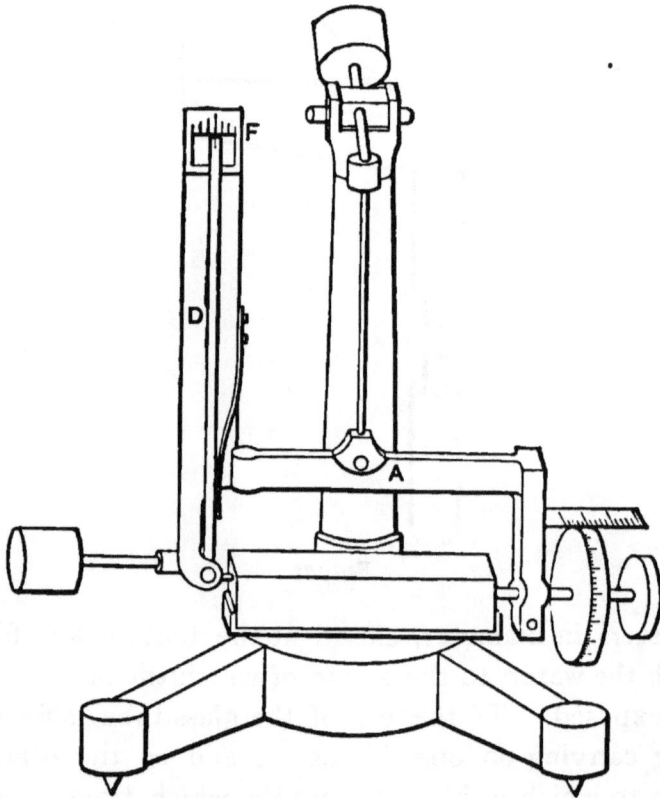

FIG. 26.

and Swiss laboratories, is an arrangement for measuring the change of length of a short bar.   It is shown
in Fig. 26.

The bars used in the test are about 100 millimeters in length and 5 square centimeters in section.

They are moulded with square cavities in the ends of the bars, in which are set small plates containing centres, to form bearings for the points of the microm-eter-screw. The change in length may be measured to 1/200 millimeter.

The measuring apparatus is suspended by a rod from a bar pivoted upon the top of the standard and balanced by counterweights so as to hang level. It consists of the caliper bar A, one end of which carries a micrometer-screw, reading to 1/200 millimeter, and the other end, a vertical bar to the bottom of which is hinged the needle D. The lower end of the needle has a point which is pressed by a spring against the end of the specimen in taking a measurement. The pressure of the micrometer against the specimen is regulated by bringing the point of the needle to zero on the scale F. A small bar of metal encased in wood is used as a standard in calibrating the instru-ment, the length of the standard being known very accurately at a definite temperature.

Le Chatelier's Apparatus.—The apparatus of Prof. Le Chatelier is designed to measure the increase in circumference of a cylindrical block of mortar. This method is recommended by the French commission for use in making hot tests, and is said to have the advantage of being much more easily manipulated than the other methods. It has also been suggested that the increase in length for long bars may not be an accurate indication of the actual swelling, as there would be a tendency for the expansion to take lines of least resistance, and therefore the transverse swelling may be more than the longitudinal.

This apparatus is shown in Fig. 27. It is described by Prof. Le Chatelier as follows:* "A much more simple and yet sufficiently precise measurement of the expansion can be made by letting the cement harden in cylindrical moulds of a diameter equal to their

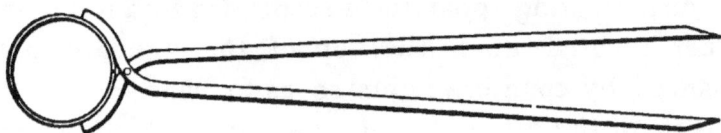

FIG. 27.

height (for example 30 mm.), constructed of metal 0.5 mm. thick, slit along generatrix and provided on each side of the slit with two long needles (150 mm. for example), which serve to magnify any widening of the slit. The widening is equal to the enlargement, not of the diameter, but of the circumference of the cylinder of cement. Very slow-setting cements or limes, the water of which would evaporate or drain away in air, it is indispensable to immerse as soon as moulded, without waiting for them to set. The immersion in water of a porous mass filled with air may sometimes, by reason of capillary phenomena, give rise to a certain expansion, and even to more or less disintegration, if the hardness be insufficient. During the moulding and until the setting has taken place the mould should be kept firm by means of a suitable holder, which is removed after setting and before the measurements are begun."

* Trans. Am. Inst. Mining Eng.

" For products of good quality, the distance between the points of the needles does not attain 1 mm. in 28 days from the time of the end of setting. This test for invariability of volume, when made cold, has but little interest, since it detects only exceptionably bad products."

ART. 54. ACCELERATED TESTS.

The fact that many cases of unsoundness in cement are not shown by the ordinary tests when extended over a short period of time has long been recognized, and many efforts have been made to find some means of determining with accuracy and within a reasonable time whether the material be reliable. The difficulty of this with a material of so variable a nature, and in which failure may be due to so many and so diverse causes, is self-evident. Each test must be directed to the determination of the presence of some particular cause of unsoundness, and all of them seem, when indiscriminately applied to all cements, to meet material which will not pass them, although of good quality, and to which they are evidently inapplicable.

Nearly all of these tests are directed to the detection of the so-called expansives, and most of them attempt to accelerate the chemical action, which causes the swelling and disintegration, by the action of heat. Some of the heat tests have proven fairly successful in use, although none have been extensively employed as tests for the reception of material.

Hot tests were first suggested by Dr. Michaelis, who proposed the use of heat to advance the hardening of

cement, with a view to determining, from the strength
gained in a short time in hot water, that which would
result in a longer period under normal conditions.
From the first experiments in this direction it appeared that the results obtained in from 1 to 7 days
in hot water might bear a definite relation to those
obtained in much longer periods at ordinary temperatures.   Later experiments showed, however, that
while this might be true of a limited class of materials,
as the composition is varied the effect of heat upon
the strength also varies in a marked degree, some
cement even showing a loss of strength in hot as compared with cold water.

In a series of tests made by M. Deval it was found
that the addition of a small percentage of quicklime
to a good Portland cement caused the cement to
attain less strength when kept in hot than in cold
water, and Prof. Le Chatelier proposed to utilize this
discovery for detecting the presence of free lime.
He suggested the use of briquettes of 1 to 3 mortar,
and the comparison of the strengths of briquettes
conserved for 3 and 7 days in water at 80° C. with
those kept for 7 and 28 days in water at ordinary
temperatures; the cements of good quality to show a
strength at least equal in hot to that attained in cold
water.

It is to be observed in considering this test that
there are some good cements which give less strength
hot than cold when of normal quality.   M. Alexandre
found that cements rich in aluminates behave in this
manner.   There are also certain cements which give
high tests in hot water notwithstanding the presence

of appreciable quantities of expansives. These are apparently silicious cements of low hydraulic index, in which the free lime while rendering the cement unsound does not cause it to lose strength in hot water during the short period of the test.

There are several methods of testing the soundness of cement by the aid of heat, which have come more or less into use, and have in many instances given satisfactory results. These all aim at the detection of the presence of expansives through accelerating their action by heat, and then observing the deformations or measuring the expansion as in the corresponding cold tests. Descriptions of these tests are given in the following articles.

ART. 55. KILN TEST.

This test was originated by Dr. Bohmé, and consists in exposing small cakes of the cement to heat in a drying oven for a definite period, and observing whether it cracks.

The specifications of the Association of German Cement Makers recommend this test as a means of forming an opinion quickly, but make the ordinary 28-day test decisive as to those cements which fail to pass the kiln test. In these specifications the kiln test is described as follows:

" For making the heat test, a stiff paste of neat cement and water is made, and from this cakes 8 cm. to 10 cm. in diameter and 1 cm. thick are formed on a smooth impermeable plate covered with blotting-paper. Two of these cakes, which are to be protected

against drying, in order to prevent drying cracks, are
placed, after the lapse of twenty-four hours, or at
least only after they have set, with their smooth surfaces on a metal plate and exposed for at least one
hour to a temperature of from 110° C. to 120° C.,
until no more water escapes. For this purpose the
drying closets in use in chemical laboratories may be
utilized. If after this treatment the cakes show no
edge cracks, the cement is to be considered in general
of constant volume. If such cracks do appear, the
cement is not to be condemned; but the results of the
decisive test with the cakes hardening on glass plates
under water must be waited for. It must, however,
be noticed that the heat test does not admit of a final
conclusion as to the constancy of volume of those
cements which contain more than 3% of calcium sulphate (gypsum) or other sulphur combinations.''

This test is considered by some authorities to be of
value for cements to be used in the air. It differs
very radically, however, from the way the material is
used in practice, as it effects the complete drying out
of the mortar. In many instances also it is very difficult to interpret, in consequence of the loss of cohesive strength due to drying, where no distortions
appear. The effect of the withdrawal of the water
necessary to the proper hardening of the mortar may
vary as the rapidity of action of the material varies.

The kiln test has sometimes been modified by using
a moist atmosphere in place of dry air. A pan of
water is placed in the oven under the specimens; the
evaporation serving to keep the air saturated with
moisture. Prof. Tetmajer used this method in a

series of comparative tests and found it to give results similar to those of his boiling test, but somewhat less effective.   His method was as follows: " The specimens are placed on a support in the oven, on the bottom of which are several millimeters of water. The heat is gradually applied so as to evaporate all the water in three to six hours,—first that which is on the bottom of the oven, then that which has been absorbed by the mortar.   Until the water is entirely evaporated the temperature remains at about 95° C. The heating is continued a half-hour after the disengagement of the vapor ceases, in such manner as to raise the temperature in the oven to 120° C.   Under these conditions the interior of the briquette will reach but little above 100° C."

" It should be remarked that by this method it is difficult to render the results comparable.   It is not possible to make the duration of heat exactly the same for all the specimens, and after the evaporation of the water the heat in the bottom is much greater than at the top of the oven."

Flame Test.—A dry heat test has been proposed, and is sometimes made in Europe, by making a ball of the cement paste about two inches in diameter and placing it on a gauze in the flame of a gas-jet.   The heat is gradually applied, so that at the end of an hour it reaches a temperature of about 90° C.   The heat is then increased until the lower part of the ball becomes red-hot, after which it is cooled and examined for cracks.   The results of this test are much like those of the dry-kiln test, and are usually difficult to interpret satisfactorily.

ART. 56.   STEAM AND HOT-WATER TEST.

This test consists in subjecting cakes of cement, prepared in the ordinary manner, to the action of steam for three or four hours, then immersing in hot water for the remainder of twenty-four hours, and examining for cracks and distortions.

Mr. Faija, by whom this test was devised, uses it in his specifications for cement in England.   He describes his method of conducting the test as follows:

" Briefly, it is a vessel containing water, the water being maintained at an even temperature of about 110° to 115° Fahr.; there is a cover to the vessel, so that above the water there is a moist atmosphere which has a temperature of about 100° Fahr.   The manner of carrying out the test is by making a pat, in the manner already described, on a small piece of glass; immediately the pat is gauged, it is placed on a rack in the upper part of the vessel, and is there acted upon by the warm vapor rising from the hot water; when the pat is set quite hard, it is taken off the rack and put bodily into the water, which, as has been already stated, is maintained at a temperature of 100° to 115° Fahr., and in the course of twenty-four hours it is taken out and examined, and if found then to be quite hard and firmly attached to the glass, the cement may be at once pronounced sound and perfectly safe to use; if, however, the pat has come off the glass and shows cracks or friability on the edges, or is much curved on the under side, it may at once be decided that the cement in its present condition is not fit for use."

Mr. Faija prefers the temperature given above, but other experimenters have seemed to get better results using a higher one. Prof. Tetmajer obtained fairly good results with a temperature just below the boiling-point—about 200° Fahr. He subjected the cakes to the action of steam for four hours, and hot water twenty hours, placing the cakes in the steam as soon as mixed.

Mr. Maclay modified this method of testing, and introduced it into the specifications of the Department of Docks, New York City. Four pats or cakes of cement prepared in the usual manner were used by Mr. Maclay for his tests, the conduct of which he describes as follows: * " One of these pats is placed in a steam-bath, temperature 195° to 200° Fahr., as soon as it is made. The second pat is placed in the same steam-bath as soon as it is set hard, and can bear the 1-pound wire. The third pat is placed in the steam-bath after double the interval has elapsed that it took the pats to set hard, counting from the time of gauging. The fourth pat is placed in the steam-bath at the end of twenty-four hours.

" The first four pats are each kept in the steam-bath three hours, then immersed in water of a temperature of about 200° Fahr. for twenty-one hours each, when they are taken out and examined. To pass this test perfectly, all four pats, after being twenty-one hours in hot water, should upon examination show

---

* Transactions, American Society of Civil Engineers, vol. XXVII. p. 412.

no swelling, cracks, nor distortions, and should adhere
to the glass plates.   The latter requirement, while it
obtains with some cements nearly free from uncombined lime, is not insisted upon, the cracking, swelling,
and distortion of the pats being much the more important features of this test.

"In hot-water tests, where the cement is very
objectionable from excess of free lime, improper burning, or other causes, the trouble generally shows itself
in the cracking or distortion of all four pats.   Where
the cement is not so bad the cracking and swelling
take place in the first three pats only, and when the
cement is still less objectionable only the first two
pats crack or swell.   The cracking or swelling of
No. 1 pat alone can generally be disregarded."

"In every case of failure and rejection the cement
should have been allowed to set hard in a normal
temperature before subjecting it to a steam-bath."

It should be noted that the effect of exposing the
cement to steam before setting seems to differ with
different material, depending perhaps upon the relative effects the heat may exercise upon the rate of
setting and upon the action of the expansives.   Where
the rapidity of setting is greatly increased by the heat,
the severity of the test may be augmented by placing
in the steam at once; but where the rate of setting is
less affected, the heat may cause the action of the
expansives to take place before the set, thus lessening
the severity of the test.   In most cases the result of
the test is the same either way, but it seems fairer to
permit the cement to set before submitting it to the
test.

Mr. Maclay, however, in his specifications does not accept the results of the steam and hot-water test as conclusive in case of failure, but only considers it as cause for suspicion of the cement failing to pass it, and adds a further test for the purpose apparently of giving the material one more chance. This test consists in testing the strength of briquettes conserved in hot water and comparing them with those kept cold.

" These briquettes are prepared and treated as follows: When making the briquettes for the ordinary cold-water tests, four additional sets of five each of neat cement, and four additional sets of mortar, one part cement and two parts sand, are prepared, and allowed to set twenty-one hours in moist air of about 60° Fahr. They are placed for three hours in a steam-bath about 195° Fahr., then immersed in water maintained at 200° Fahr., after which they are broken when two, three, four, and seven days old respectively, and the breakings compared with the normal breakings of briquettes seven and twenty-eight days old kept in cold water."

" The writer finds, in a general way, that the averages of the breakings of hot-water briquettes of pure cement, four days old, are nearly as high as the normal seven-day breakings cold, and the hot water seven-day breakings of the pure cement are nearly as high as the normal twenty-eight-day breakings cold, where the cement is of good quality. Where the cement is poor, and the pats show cracking and distortion, there is generally a remarkable falling off in the strength of the hot-water briquettes from the above comparison.

and one system can therefore be used as a check on the other.''

This is practically the same test mentioned in Art. 54, as proposed by M. Le Chatelier. Its use in this manner is recommended by M. Candlot:

'' The cements which contain free lime show less resistance in hot water than in cold. The cements of good quality in hot water show resistances equal to or greater than those in cold. Cements properly proportioned and homogeneous, but not completely burned, give, in this test, satisfactory results.''

As already pointed out, however, the relative strengths hot and cold do not depend altogether upon the presence of expansives, and it is questionable if this method is as accurate as that it is designed to check. Some unsound materials certainly give high results in the measurement of the strength of briquettes conserved hot, while there seems to be no authentic instance of any unsound cement being accepted on the steam and hot-water test, although good cement may perhaps be condemned.

ART. 57.   BOILING TEST.

The boiling test, which is very similar in effect to that by steam and hot water, was first suggested by Prof. Tetmajer of Zurich. It consists in placing the mortar to be tested in cold water, and then gradually raising the temperature of the water to boiling. Prof. Tetmajer's method is to place the pats in cold water, immediately after gauging, raise the temperature to boiling in about an hour, continue boiling for three hours, and then examine the pats for checking and softening.

This method seems rather more severe in its effects upon the mortar than the other hot tests, although in general differing but little from the steam and hot-water test when a boiling temperature is employed in that test; the action, however, seems to be more energetic, and less time is required to arrive at the same results.

It seems desirable in using the boiling test to permit the cement to set before subjecting it to the test, as giving a more reliable indication of value. The results of the test is in most cases practically the same whether the cement has previously set or not. When cement is subjected to the boiling test before setting takes place it is necessary to exercise much care in the manipulation of the test to avoid any disturbance of the mortar through the motion of the water when heated. The results of the test also depend somewhat upon the rate of setting of the material, and upon the influence of heat upon the rate of setting. With quick-setting cements this action is unimportant, but with those very slow the heat may cause the action of the expansives to take place in advance of the setting, or the cement may remain soft until late in the test, and appear to fail, in consequence of disturbance due to the ebullition of the water.

Prof. Tetmajer recommends for this, and in fact for all pat tests, that the cakes shall not be made with thin edges. His method is to roll a ball of the mortar, and then flatten the ball to the required thickness. The consistency of the mortar is determined by the requirement that it shall not crack in flattening or run at the edges.   For tests in boiling water this seems desirable, but for pats to be used in the ordinary cold tests the thin edges are of advantage in expediting the results where unsoundness exists in the mortar.

The boiling test is frequently used in connection with apparatus for measuring expansion, in place of observing the distortions or cracks.   The Bauschinger caliper apparatus is sometimes employed in this way, the bars being subjected to the boiling test, and the increase in length noted.   The Le Chatelier apparatus (Fig. 27) is also usually employed in this manner. The French commission upon methods of testing materials recommend the use of the Le Chatelier apparatus for this purpose, in addition to the cold test given in Art. 52, as follows:

" *Hot Tests.*—For these tests cylindrical test-pieces are employed, 3 centimeters in diameter and 3 centi-meters high, made in metal moulds 1/2 millimeter thick, cut on a generatrix, and carrying, one on each side of the slit, two needles, 15 centimeters long.   The increase of the distance between the ends of the needles gives a measure of the swelling.

" The molds as soon as they are filled are im-mersed in cold water.   After an interval of not more than twenty-four hours beyond the completion of the set the temperature of the water is gradually raised to 100° C. in from a quarter to half an hour.   The temperature is maintained at 100° for six hours, and then it is allowed to cool before taking the final measurements."

" This method of testing is not applicable to quick-setting cements."

" The standard test for deformation is to be made upon neat cement of standard consistency."

The boiling test is more simple than the steam and hot-water test, and requires very little in the way of apparatus. It may readily be made anywhere without difficulty. In a laboratory where apparatus may be kept in continuous operation the steam and hot-water test offers some advantages in ease of operation, and in permitting continuous study of material under examination beyond the limits of the regular test.

ART. 58. PRESSURE TEST.

Dr. Erdmenger has devised a high-pressure steam test for soundness. In this test the pats are allowed to harden for three days, and are then exposed to steam for six hours at a pressure of from 3 to 20 atmospheres. The originator has used this test for a number of years, and claims that it gives very satisfactory results, and that when properly carried out it enables a complete and rapid judgment to be formed on a cement containing magnesia.*

As with all accelerated tests, there has been much dispute as to the reliability of this one, some authorities claiming that many of the best Portland cements fail under it, others considering it nearly infallible. It is a test requiring more extensive appliances for its execution than the other hot tests, and has not been so largely used as the others.

* Journal Society of Chemical Industry, vol. XII. p. 927.

The test is executed both upon the neat cement
and sand-mortar, the severity of the test being greater
upon the neat cements.   Dr.   Erdmenger claims that
the best cements show no defect under this test at
high pressures (forty atmospheres); that others may
show defects at high pressures, although safe in practice (especially in the neat test), but they are not first
quality; while cement which cannot stand the pressure
test of about twelve atmospheres in the sand tests
should be rejected as faulty.

ART. 59.   CHLORIDE-OF-CALCIUM TEST.

The action of chloride of calcium in the water used
in gauging or conserving the briquettes is similar to
that of heat.   The results of the experiments of
M. Candlot have already been discussed (Art. 22).
In presence of calcium chloride the slaking of lime is
greatly accelerated, the effect increasing as the solution is more concentrated.   It also makes the setting
slower, especially for cement rich in aluminate of lime.

If cement containing free lime be gauged with a
concentrated solution of calcium chloride, it may
therefore cause the lime to become slaked before
setting, so that no subsequent swelling will take place.

If the solution be less concentrated, the lime may
not be completely slaked before setting unless the
quantity be very small, while the slaking is hastened
and swelling occurs after the setting if the free lime
be in sufficient quantity.

If a chloride-of-calcium solution is used to keep the
cement after setting, the expansive action of the free

lime is accelerated, and a very small quantity may produce swelling.

This test, as devised by M. Candlot, is to mix the cement paste with a feeble solution of calcium chloride, which causes the slaking before setting of so small a percentage of lime as may not be objectionable in the cement, but does not eliminate a larger quantity. Then immerse the mortar in the same solution, and thus augment the swelling if the free lime be present in serious quantity.

The method of conducting the test as proposed is to " mix the paste for the cakes with a solution of 40 grammes of chloride of calcium per liter of water, allow them to set, immerse in the same solution for twenty-four hours; then examine for checking and softening as in the other tests."

The chloride of calcium is said to exert no action upon free magnesia. Upon free lime it seems, however, to be quite efficient, and it may often be useful in pointing out doubtful material.

The action becomes less energetic as the temperature is increased, and it should not be used in connection with the hot tests.

ART. 60. VALUE OF THE ACCELERATED TESTS.

The reliability of the various accelerated tests for determining the soundness of cement in use is a matter concerning which there is much dispute amongst authorities on the subject. These tests have not as yet come into general use, and considerable opposition has been developed to them, although in certain instances they are employed.

The question of adopting the accelerated tests was discussed by the Association of German Cement Manufacturers in 1891, and it was voted to adhere to the standard test already existing. It was also stated in the report that experiment had not proved the inaccuracy of the standard test (the twenty-eight-day cold-pat test), and while the accelerated tests may be useful to the manufacturers in determining the character of their cements, they should not be used by the consumer, with a view to forming a sound opinion on the constancy of volume.

The results of a number of experiments are also given in this discussion, showing that a number of cements which had not withstood the *kiln* test were sound when used under water at normal temperatures, or if placed in air after being kept moist for several days. It may be remarked also that most of the discussion seemed to refer mainly to the dry-heat test.

Some of the leading German cement experts, however, are strong advocates of the use of heat tests. Dr. Michaelis expressed his approval of them, and stated that he had experimented with and used them satisfactorily for a number of years at the Charlottenburg experiment station. Dr. Erdmenger also declares that experience has shown the high-pressure steam test to give an accurate determination of the permanence of volume of the material. He states that most of the best German cements have stood this test up to a pressure of forty atmospheres for sand-mortars and several for neat cement, the best of them showing no defect whatever; and he thinks there is no ground for the statement that many good cements will not pass the test.

Prof. Bauschinger found that cements which had given good results in the ordinary 28-day test, and also upon the cold-pat test for a year, failed when formed into prisms of 1 to 3 mortar for testing in the Bauschinger caliper apparatus. Expansion was detected in six months by measurement, and afterward became visible to the naked eye.

The German Conference upon Methods of Testing * recommend the continuance of the present practice (the kiln test for quick determinations and the 28-day cold-pat test as decisive), but they state that " the boiling test may undoubtedly be considered as the most conclusive and rapid test for the determination of constancy of volume of Portland cement, of slag-cement, and of trass," and refer a particular test to the sub-committee for further examination and report.

The French Commission upon Methods of Testing † recommend the use of the boiling test as the best method of arriving at a quick determination of the. permanence of volume for Portland cement, the amount of the increase in volume to be measured directly, instead of simply observing the effect upon a small pat.

---

* Resolutions of the Conventions held at Munich, Dresden, Berlin, and Vienna, for the purpose of adopting uniform methods for testing materials, by J. Bauschinger; translated by O. M. Carter and E. A. Gieseler (Washington, 1896).

† Commission des Méthodes d'Essai des Matériaux de Construction; Rapport (Paris, 1894 and 1895).

In England the steam and hot-water test has been introduced by Mr. Faija, using a low temperature; while in the United States the same test is employed in the specifications of the New York Department of Docks, using temperature near boiling. In some instances the boiling test has also been used in the United States by manufacturers in the study of their product, with very good results.

The advisability of adopting some form of accelerated test in ordinary specifications is still an open question, needing for its determination more accurate knowledge than is now available of the behavior of the various kinds of cement when subjected to such tests.

Numerous experiments have been made for the purpose of deciding the matter, the results of which differ so widely from each other as to involve the question in great confusion. Most of these experiments have been made by studying the action of heat upon the various ingredients to which unsoundness is usually attributed, and arguing from the results whether the hot test gives an accurate indication of the presence of these ingredients in the cement. The most common method has been to mix a small quantity of quicklime with a good Portland cement, and then observe the action of the test upon the mixture. This involves the assumption that certain percentages of free lime are sufficient to render the cement unsound. The results of tests made in this manner are also subject to much variation, due to the nature of the cement, its rapidity of action, and the quantity of free lime which may have been originally present in it.

These experiments have been of much value in showing the effect of the accelerated tests upon various substances, and in discovering the reasons for many of the apparently contradictory results with them. The question of the ultimate adoption of tests of this character must, however, be determined by a comparison of the results obtained by their use upon the material as found in market with the action of the same material in practical use. To this end experiments are desirable which shall systematically compare the results of accelerated tests upon ordinary cements with the results of tests under normal conditions extended over long periods of time. An extended series of experiments of this character has already been carried out by Prof. Tetmajer at Zurich,* and a smaller series upon material in use in the United States at the laboratory of the College of Civil Engineering at Cornell University.

A careful study of the available results of experiment seems to justify the following statements:

1. Small percentages of uncombined lime or magnesia in the cement are commonly detected by the use of the heat tests, and the same ingredients in sufficient quantity render the cement unsound in ordinary use.

2. Cement liable to change of volume when employed under normal conditions is almost invariably detected when submitted to the hot-water test. There seems to be no well-authenticated instance of failure to condemn really defective material.

---

* Methoden und Resultate der Prüfung hydraulischen Bindemittel (Zurich, 1893).

3. Nearly, if not quite, all of the best brands of Portland cement, and many of natural cement, as found in market, readily meet the requirements of these tests, which therefore do not impose so severe limitations upon the choice of cement as is commonly supposed. With natural cement the results of these tests vary somewhat with the character of the cement, and the same tests do not seem to be universally applicable. This, however, is a matter which can only be determined by careful experiment upon each of the various classes of natural cement. Many of them bear the severe tests fully as well as the Portland.

4. While these tests rarely, if ever, fail to detect an unsound cement, and most good cements readily pass them, there are occasional instances of cements condemned by the heat tests, which are not unsound when kept at normal temperatures in fresh water. These cements for the most part seem to fail if kept in dry air or are subjected to the action of sea-water. Apparently, a degree of unsoundness may exist which is sufficient to cause the change of volume to take place in a short time in hot water or in a longer time in dry air, while in cold water the action of the expansives takes place without injury to the mortar. This is shown by the fact that pats of mortar which had failed in the boiling test at the time of mixing, after being kept several months in cold water, and then subjected to the boiling test, were found to stand the test perfectly, showing that the action of the expansives must meanwhile have taken place.   In

several instances pats of cement acting in this manner were found to blow in dry air.

The unsoundness of cement condemned by the heat tests, when the mortar is to be kept submerged in fresh water, is therefore, in many instances at least, questionable. In the large series of tests made by Prof. Tetmajer, out of 139 samples of Portland cement 17 failed in the boiling test, all of which also failed in a long time in dry air, while only 2 were defective in long time in fresh cold water. In the tests made by the author, the percentage of failures to total number of samples is less than at Zurich, 5 in 53, and 3 of the 5 samples which were rejected by the hot test were later disintegrated when kept in cold fresh water.

5. The fineness of the cement has an important bearing upon its behavior under the heat tests. All ordinary cement probably contains some small proportion of expansives, which in the finely ground material may become hydrated without injury to the mortar; but if the fine material be sifted out, and only the coarse particles employed in the test, the slower action in the larger particles may cause distortion and cracking to take place. Whether the same difference usually follows in cold water has not been satisfactorily determined.

6. Portland cement which contains free lime in sufficient quantity to cause it to swell in the hot test may frequently be made to pass that test by adding a small quantity of sulphate of lime. The action of this salt upon the setting of cement has already been discussed (Art. 21). The rules of the Association of

German Cement Makers permit the addition of a small proportion of sulphate of lime for the purpose of regulating the rate of setting, and conclude that no injury is thereby caused to the cement.

The action of the sulphate of lime to correct the expansive action of free lime is but imperfectly known. The fact that it causes cement containing free lime to pass the hot tests is well known, but whether the corrective influence extends to the action of the expansives when the material is used under normal conditions has not as yet been satisfactorily determined.

The strongest objection that has been urged against the use of hot tests is that they fail to detect free lime in the presence of the calcium sulphate. The justice of this objection can only be decided by experiments extending over considerable periods of time to determine whether the material so passed is sound under normal conditions. Doubtless for use in sea-water it would be necessary to limit the quantity of sulphuric acid.

The whole subject of hot tests must still be regarded as in the experimental stage, and further experiment is necessary to determine more fully the connection between the results of tests and the action of the cement in use.

Under present conditions it may be said that the presumption is fairly against the soundness of a cement failing to pass the hot-water test and in favor of that which succeeds in passing it; but variations in the cement and in the conditions under which it is employed may affect the results, and they cannot be relied upon with certainty for all material. Upon ordinary work to be kept in fresh water there is probably no considerable danger of unsoundness in the use of any good brand of cement, although failure sometimes occurs; but for the more important works, and particularly those to be subjected to the action of sea-water, it is reasonable to apply such tests as are likely to insure good material, even at the risk of excluding other good material.

ART. 61.   AIR-SLAKING.

Sometimes fresh cement, when first opened after being shipped, will, if tested at once, show an abnormally rapid rate of setting, and subsequently harden very slowly, so that on short-time tests very low tensile strength may be given. If, however, this cement be exposed to the air for a few days, it may resume its natural rate of setting, and attain proper strength upon the tests. In some laboratories it is customary to thus expose cement to the air a short time before testing, and this process is termed *air-slaking*.

The propriety of air-slaking in testing cement is questioned by some engineers, upon the ground that the cement to be used in the work will not be treated in the same manner. In England it is customary to give such exposure to all cement to be used upon important work for at least ten days, but in the United States the cement is commonly used just as received from the manufacturer.

The general practice seems to favor air-slaking in testing, and probably a cement capable of regaining its normal condition in a few days' exposure will not endanger the work, even if used at once, but it would doubtless be better in using such cement to air-slake the whole before using.   It may be remarked, however, that air-slaking does not ordinarily seem neces· sary.   The cement commonly placed upon the market by the best makers does not need it.   While it may be allowable to give the material the benefit of the operation, probably few instances of rejection would occur on account of its omission.

In many instances the effect of air-slaking a cement requiring it disappears with time; that is, the strength of the mortar after three or six months may be as great for that mixed before as for that mixed after air-slaking, although the difference of strength on a test extending over a few days is very considerable.   This would indicate the necessity of air-slaking the whole of the material if early strength is to be developed in the work, although the ultimate strength might perhaps be satisfactory either way.

If the cement blows or shows unsoundness on the first test, it should not be used without exposure, as it would indicate a degree of unsoundness likely to be serious, even though this also disappear in the second test.

The question is simply as to the quantity of the expansives which may be present without danger to the work in which it is used.

CHAPTER VIII.

SPECIAL TESTS.

ART. 62.   TESTS OF ADHESIVE STRENGTH.

THE ability of cement-mortar to firmly adhere to any surface with which it may be placed in contact is one of its most valuable properties, and quite as important as the development of cohesive strength. Tests for adhesive strength are not commonly employed as a measure of the quality of the material, because of the uncertain character of the test and the difficulty of so conducting it as to make it a reliable indication of value. The adhesive properties of the cement are to a certain extent called into play in the tensile tests of sand-mortar, and may be inferred from a comparison of neat and sand tests.

Adhesive strength is developed more slowly than that of cohesion. The difference between the two, which is usually considerable during the early period of hardening, is gradually lessened with time. This is illustrated in the greater time required for sand-mortar than for neat cement to harden.

Experiments upon the adhesion of mortars to various substances are sometimes made, both for the purpose of comparing various cements or methods of use, and to study the relative effects of various kinds

of surfaces.   Such experiments are quite desirable
with a view to the extension of knowledge of this very
important quality.

The common method of making this test is to prepare briquettes, of which one half are of neat cement
or sand-mortar of the ordinary form for tensile specimens, and the other half a block of stone, glass, or
other material to be used, of the same section as the
mold at its middle, and arranged to be held by a
special clip in the testing-machine at the other.

In Germany and Switzerland the apparatus shown in
Fig. 28 is employed.   The test-piece is shown at a;

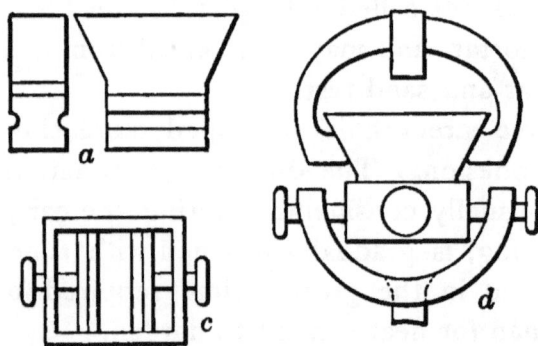

FIG. 28.

it has a section of 10 square centimeters—twice that
of the standard tensile specimen.   The mortar end is
enlarged to a wedge shape to catch the upper clip of
the testing-machine, while the other end is formed of
a parallelopiped of marble or ground glass for standard
tests, with a cylindrical groove cut in its side, which
fits into a special clamp (shown at c).   This clamp is
held by screws in the lower clip of the testing-machine,
as shown at d.   For forming the briquettes molds

are used, in the bottom of which the blocks are placed and the mortar filled in on top.

In some laboratories blocks of marble have been used for standard tests in comparing different materials; but the large amount of labor involved in the preparation of the blocks, and the difficulty of getting always the same surface, has been a bar to the extension of this method. Ground glass has been more commonly employed, the same blocks being repeatedly used. Dr. Michaelis has also used for this purpose standard blocks of cement-mortar, of the same form, which are easily prepared, and more uniform in material and surface.

The German Conference on Methods of Testing, before mentioned, did not define a standard test, but referred the matter to a sub-committee, with the recommendation that the German apparatus just described be utilized.

The French commission recommend the use of a briquette of double T form, suggested by M. Candlot. It is shown in Fig. 29. A mold is employed, made to the form of half the briquette, which is set down over the block to be used in forming the specimen.

The recommendations of the commission are as follows:

" A. To compare the adhesive strengths of cements, there is submitted to the tension test a briquette of the double T form, shown above, each of the materials to be studied constituting half of one of the specimens.

" B. Standard tests, intended to compare the adhesive strengths of various cements to the same material.

" (a) The standard adhesion-blocks are prepared of mortar composed of one part, by weight, of artificial Portland cement, passed through the sieve of 900 meshes per square centimeter, and two parts of standard sand No. 3 (that passing a sieve of 1 mm. open-

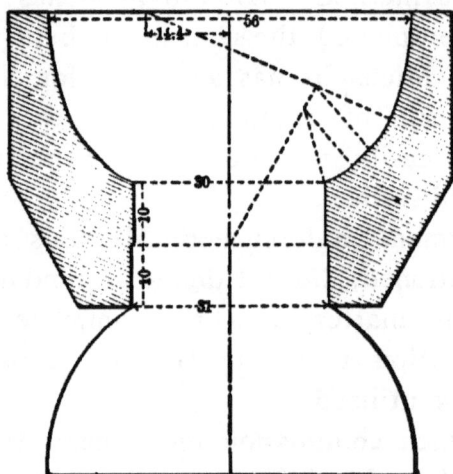

FIG. 29.

ings and retained upon one of 1/2 mm. openings). The mortar is gauged with 9 per cent water and strongly compressed in the moulds, in the bottom of which is placed a movable metallic block. The adhesion-blocks are immersed in fresh water after twenty-four hours, and kept thus until used, or at least for twenty-eight days. When they are to be utilized, they are dried and the surface is smoothed with emery paper.

" (b) The standard plastic mortar (see Art. 44) is employed for these tests, introduced by the pressure

of the trowel into the moulds, placed in such manner
that the standard adhesion-block forms the base.
The briquette is removed from the mould when completely set."

" (c) The briquettes are tested in the same manner
and at the same periods prescribed for tensile tests."

" C. Tests intended to compare the adhesive
strength of a cement to various materials;

" (a) For these tests the same methods are adopted,
with the difference that the standard blocks are replaced by blocks formed of the various materials to be
used. If the material can be moulded the block may
be made in the same way as the standard block. If
the material is solid, a plate several millimeters thick
may be formed having one face dressed. This is
placed in the bottom of the mould and the block filled
out with neat cement."

ART. 63.   CHEMICAL ANALYSIS.

Chemical analysis is of very great value in the study
of the properties of various cements, and is commonly employed by manufacturers in regulating the
quality of their products. It is not commonly used
for the purpose of determining the quality of a cement,
and is not of much value as a test for the reception of
material.

The quality of the cement depends not only upon
the ingredients being properly proportioned, but also
upon the state of combination of the ingredients, and
this in turn depends upon the manipulation given the
material in manufacture. Analysis shows the propor-

tions of the various ingredients, but does not show
their state of combination.   The results of an analysis
may show that the composition is such that a good
cement may be made from the ingredients, but other
tests are necessary to show whether it has been made.
Chemical analysis may, however, prove a cement to
be bad through its containing objectionable proportions of ingredients known to be injurious.   Thus the
allowable percentage of sulphur compounds is sometimes prescribed for cement to be used in maritime
work; in fresh water they may be unobjectionable.
The expansive elements, free lime or free magnesia,
cannot be detected by analysis, as their presence is
not necessarily dependent upon the proportions of
total lime or magnesia present.

The French specifications, devised by M. Guillain,
for Portland cement reject those containing more than
1% of sulphuric acid, or sulphides in appreciable quantity; while those containing more than 4% of oxide of
iron, or which give less than 44/100 for the ratio of
the sum of the weights of silica and alumina to the
weight of the lime are to be regarded with suspicion.

A large percentage of volatile elements in a cement
indicate either insufficient burning or deterioration
with age through exposure to the air.

M. Candlot states that a chemical analysis may be
useful in showing the adulteration of cement, sometimes practised in Europe.   Upon sifting the cement
and separately analyzing the coarse and fine portions
an unadulterated cement should show practically
identical results for the two analyses.   He also states
that blast-furnace slag, which is a common adultera-

tion in Portland cement, may sometimes be discovered by the odor of sulphuretted hydrogen upon treating it with hydrochloric acid.

Method of Analysis.—The following method of making an analysis of cement is given by M. Durand-Claye:*

" Weigh two grammes of the specimen in powder, place in a porcelain evaporating-dish with a little water and hydrochloric acid in excess. Evaporate dry in a sand-bath at a moderate temperature. The mass should be stirred several times with a glass rod after the silica has jellied. When the drying is complete, place the dish in a furnace and heat quickly, but not to redness, to remove the last trace of acid and humidity. After cooling, treat the new material with a slight excess of hydrochloric acid, and evaporate a second time to dryness. Then treat again with acidulated water. The liquor should appear yellow, and the residue white, not reddish. If it has that color, let the dish and its contents heat some time in a sand-bath until all the red color disappears.

" *Silica.*—Place the contents of the dish upon a filter. The white powder obtained is the silica of the specimen. Dry, calcine, and weigh in the ordinary manner.

" *Silicious Sand.*—The cement sometimes contains grains of inert silicious sand, which is found with the silica. It is recognized by the sound of the grains on the bottom of the dish when stirred with the glass rod. To determine the silicious sand, dissolve the cement

---

* Chimie appliquée à l'art de l'ingénieur (Paris, 1885).

in hydrochloric acid, and let the liquor digest during one day.   Then, upon decanting, the silicious grains rest upon the bottom; wash, collect, and weigh.

"*Alumina and Iron.*—The filtrate, after removing the silica, is treated with a few drops of nitric acid, and brought to ebullition for a moment.   The protoxide of iron, if any, is transformed to peroxide. Then, by saturating with a slight excess of ammonia, and boiling, a precipitate is formed, which is collected on a filter placed in a funnel above a flask.   It is washed carefully with warm water. When well drained, it is placed on a sand-bath, dried, burned, and weighed. The weight found, diminished by that of the cinders of the filter, is that of the peroxide of iron and alumina soluble in acids.   The ammonia precipitates the two bases from the solution, while it is without effect upon the salts of lime, and does not affect the salts of magnesia previously acidulated.   When the limestone contains much magnesia, like dolomite, some of that base may be precipitated by the ammonia. In this case it is prudent to redissolve the precipitate in hydrochloric acid, after having separated by filtration, and to reproduce it a second time with ammonia.

"*Lime.*—Add oxalate of ammonia to the last filtrate, and the lime only is precipitated, as magnesia is not separated in the water containing ammonia salts. In order that the oxalate of lime may be well precipitated, it is necessary to employ a large excess of the reagent.   The precipitate is collected better when the temperature is higher; it is therefore well to heat the liquor, and dissolve the oxalate of ammonia in boiling water.   In order to obtain a complete precipitation the liquor must be alkaline. The precipitate should not be filtered immediately, as the liquor continues working. The flask should be put in a furnace and boiled for half an hour, or still better be left in a sand-bath half a day. The boiling in the furnace of a grainy precipitate like oxalate of lime gives rise frequently to shocks, which perhaps throws the material out. When the precipitate is well formed and the liquor clear, place upon a filter in a funnel over a flask and wash carefully with hot water; then, placing the funnel in a sand-bath, dry the filter and precipitate before calcining. A simple scorching is not sufficient. The oxalate at the temperature of burning is decomposed more or less completely: a part becomes caustic lime; some is transformed to carbonate, or it remains as oxalate. It is therefore necessary to transform it to a definite compound.

" When the means are at hand for a sufficiently energetic burning to heat the platinum cherry-red, put the precipitate, with the filter, in a platinum crucible furnished with a cover, and calcine. Five minutes are sufficient with an annealing bellows. The oxalate is transformed to caustic lime, which is immediately weighed in the crucible when cold. · The weight, less cinders of filter, is the lime contained in the sample. It is well to assure that the calcination is complete; for this purpose, after weighing, pour a little water in the crucible, then acid; if any carbonate is present, bubbles of gas will be given off.

" When an apparatus for heating so energetically is lacking, the oxalate is sometimes changed to carbonate. Place the precipitate in a platinum dish and burn in a muffle, or to a heat which makes the dish red. When it is cold, pour a solution of carbonate of ammonia; then evaporate in a sand-bath, dry over a furnace, and weigh. It is well to renew the treatment of carbonate of ammonia until two weights are obtained which are in accord. The weight of lime is 0.56 that of the carbonate.

" *Magnesia.*—The filtrate contains the magnesia, with the ammonia salts which are introduced in the course of the analysis. Precipitate with phosphate of soda. Let it stand cold half a day before filtering, until finally assured that the ammonia-magnesian phosphate which is formed is entirely crystallized, or at least that not more than insensible traces remain in solution. When placed on the filter some of the crystals adhere to the flask, and may not be detached by washing. These are dissolved in acid and then again precipitated with ammonia in non-adherent granular crystals, which readily wash upon the filter. The washing from the flask to the filter should be done with cold water charged with ammonia. Pure water, especially if warm, dissolves an appreciable quantity of the precipitate. After drying, burn with the filter and weigh. 40/111 of the weight, less the cinders of the filter, gives the weight of magnesia.

" *Volatiles.*—Weigh a new specimen in a crucible and calcine quickly for five or ten minutes. The crucible when cold is weighed with its contents, which are composed of lime and the other fixed elements. The difference in weight is the loss upon ignition.

" *Sulphuric Acid.*—If it is desired only to obtain the sulphate of lime, take a specimen (5 grammes for example) and digest cold in a solution of carbonate of ammonia for twenty-four hours. Agitate the mixture often, especially in the beginning. The sulphate of lime is decomposed, carbonate of lime and sulphate of ammonia being produced. Filter, saturate the liquor with hydrochloric acid, and boil to remove all carbonic acid. Then pour in chloride of barium, and the sulphuric acid is precipitated as sulphate of barium, which is collected on a filter, dried, and weighed. The filtration should not take place immediately: it should stand until the sulphate of barium gathers and the liquor clears.

" Sulphides.—If the total sulphur is to be obtained, it is to be brought into contact with an oxidant, which transforms the sulphide into sulphate. A little chlorate of potassium is added, and then it is attacked with precaution by hydrochloric acid. Evaporate dry, and digest the residue, after cooling, with carbonate of ammonia. The analysis is then like the preceding. If the two methods be simultaneously applied, and by the second a result be found larger than by the first, the difference represents the sulphuric acid formed from the sulphides.

" Alkali.—Take a special specimen and separate the silica, alumina, and iron as already indicated. Pour the filtrate so obtained in a porcelain dish and evaporate dry, then calcine in a muffle to drive off the ammonia salts and decompose the oxalates. The residue is composed of alkaline salts, and of magnesia, free or carbonated. The latter are insoluble, and by treating with water and filtering, a liquor is obtained containing only the alkaline salts. The liquor is evaporated to dryness in a platinum dish in presence of an excess of sulphuric acid, which displaces the other acids. The residue is alkaline sulphate. Heat to redness to drive off all excess of acid, and weigh. Treat with water acidulated with hydrochloric acid, and precipitate the sulphuric acid with chloride of barium. The weight of sulphate of barium gives that of sulphuric acid, which subtracted from that of the alkaline sulphates gives fixed alkali.

" When the specimen contains considerable sulphuric acid this method may give erroneous results. The sulphate of magnesia is only partially decomposed at the moment of calcination of the ammonia salts, and a part of that salt is found with the alkaline sulphates. To remove that cause of error, after having eliminated the ammonia salts and treated the residue with water, pour on a little baryta water, which precipitates the magnesia and gives at the same time compounds insoluble in sulphuric or carbonic acid. The filtrate contains then only alkaline chlorides, free alkali, and baryta. Saturate with sulphuric acid and filter, then evaporate dry, and calcine, and obtain the residue of alkaline sulphates as in the other case.

" *Separation of Iron and Alumina.*—After having burned and weighed these two bases, digest and heat with hydrochloric acid until dissolved. Mix with citric or tartaric acid until ammonia in excess does not produce a precipitate; then add the sulphohydrate of ammonia. The alumina is not precipitated, while the iron passes to the state of insoluble sulphide. Boil and collect on a filter, dry, burn, and weigh. The sulphide decomposes and becomes peroxide.

The alumina is the difference between the total weight and that of the peroxide of iron.

'' Or, after dissolving in hydrochloric acid, evaporate the larger part of the hydrochloric liquid so as to have a slight excess of acid; then pour a solution of pure potash in excess, and boil until the precipitate obtained presents a dark reddish brown color. Filter and wash the precipitate of peroxide of iron. An appreciable quantity of alumina will be precipitated with the iron; to remove this, redissolve in hydrochloric acid and repeat the operation a second or third time. The absence of alumina is shown by the ammonia salt giving no precipitate with the filtrate.

'' This operation requires great care and considerable time.''

ART. 64. TESTS FOR HOMOGENEITY.

In Europe various tests have been proposed for the purpose of detecting the adulteration of Portland cement. These tests are not usually of a nature intended for use in specifications for the reception of material, but may sometimes be of use in studying the characteristics of various brands of cement or in classifying a product of doubtful character. The materials sometimes met with as adulterations include powdered limestone or shale, blast-furnace slag, hydraulic lime, trass, etc. The nature of the tests depends upon that of the adulteration to be discovered.

The specific-gravity test is sometimes utilized for this purpose, the foreign matter being lighter than cement. The differences, however, are so small that the test at best is a rather uncertain one.

The determination of the loss upon ignition may show the presence of foreign matter containing volatile elements, while chemical analysis, separating the fine and coarse parts of the cement and comparing the results, is sometimes resorted to, as stated in Art. 63.

Microscopic Test.—The use of the microscope for the purpose of determining the character of the substances present in cement has often been proposed. Prof. Le Chatelier made a careful study of Portland cement by examining sections of unground clinker by polariscopic analysis. He thus demonstrated the possibilities of this method in the scientific investigation of the nature of the material.

This method is not, however, applicable as a test for homogeneity. Attempts have also been made to determine the character of the cement by studying the grains of which the powder is composed under the microscope. For Portland cement, it has been observed that the active portion of the cement is composed of grains of angular form and metallic lustre, and that the parts of earthy appearance are probably inert. It has also been found that the color of the grains seems to bear some relation to their value in the cement. Further study may reveal more positive indications of value based upon the microscopic appearance, but it seems unlikely that this method will be applied to the determination of value in practice.

For tests of homogeneity, the employment of an ordinary magnifying-glass may perhaps give useful results. M. Feret, who made a study of the matter and presented a report to the French Commission upon Methods of Testing Materials, recommends that the material to be examined be sifted through the sieve of 4900 meshes per square centimeter and the portion retained by that sieve be used in the examination, on account of the difficulty of observing the grains when mixed with the impalpable powder. It is also suggested that two glasses be used: the first of a power of about three diameters to examine the general appearance and uniformity of the material; the second of about eight diameters to study in detail the various grains. The material is placed upon a black surface to be examined. The points to be noticed are the form, color, and transparence of the grain and the appearance of its fracture. It is also desirable to test the hardness with a steel point and the solubility in a drop of water or acid. It is to be borne in mind that the character of the material in the coarse particles is not necessarily the same as in the finer portions, or at least the proportions of the various ingredients may not be the same.

M. Feret found that pure cement, thoroughly well burned, gave "grains all of the same appearance, black, opaque, angular, hard, and with a rough fracture. Mixed with a drop of water they show at first no change, but after several minutes a sort of halo appears, produced by the beginning of crystallization of the soluble compounds. These grains color immediately in hydrochloric acid to a yellow, but are completely dissolved with difficulty.

"Rock less burned gives grains equally opaque, but of color less deep, varying to brown, gray, or green. The underburned rock gives gray or yellow grains, which crush easily under the point of a knife, and are attacked by acids with the disengagement of carbonic acid."

" When, instead of material prepared in the laboratory and thoroughly homogeneous, the residue obtained from market grades of cement is used, the appearance is very different even if the material be of best quality. The color is less deep and the material appears very heterogeneous. Such cements contain always, with many of these black, brown, or green grains, a large number of more vitreous material, green, yellow, and white, without containing foreign matter."

" When the cement is less well burned the color of the large grains becomes more clear, and the glass shows an increasing number of gray friable grains, the black grains decreasing in number."

" It is not uncommon to find in the cement clinker pieces of bright colors, blue, green, violet, red, and white; these materials are usually soft and porous."

" Carbon in black and brilliant grains, with conchoidal fracture, is found in all the specimens, and easily known."

" The débris of flint from the millstones is not easily distinguished from certain underburned cement grains; they are the gray morsels, hard and opaque. They differ from the cement grains in not being attacked by water or acids."

" The appearance of grains of slag vary according to the nature of the slag. Commonly, the grains are compact, of a bluish gray color, and smooth clean fracture. Sometimes the grains are vitreous and black. The granular slags employed in making slag-cements are soft, and leave few grains. Their débris has the appearance of colorless glass, or tint of yellow or green, and fractures easily under the point of a knife.''

'' Grappiers give round grains, usually more clear in color than the grains of cement.''

'' Gypsum added to cement clinker before grinding may show large white crystalline grains easily seen. They may be identified by their hardness and solubility. Plaster is difficult to recognize, because it grinds fine and leaves no crystals.''

When plaster has been added to cement, it may sometimes be detected by separate analyses of the fine and coarse parts, as to sulphuric acid. The fine parts containing the plaster give the higher results.

Le Chatelier's Test.—This method of detecting adulteration, proposed by Prof. Le Chatelier, consists in forming a liquid, by the mixture of methylene iodide and benzine, of density slightly below that of cement and above that of slag, and separating by its means the cement from the adulteration. It is thus described by Prof. Le Chatelier:

'' The first operation is the preparation of a liquid of proper density for the separation,—2.95 for example. In this liquid the cement sinks to the bottom and the slag floats on the surface. To prepare the liquid add to the methylene iodide of density 3.1 a small quantity of benzine, stopping the moment a crystal of aragonite of density 2.94 just remains at the surface. It is well not to make the mixture directly, but to make two mixtures—the one a little lighter, the other a little heavier, than the density sought;

thus obtaining a more progressive variation, and more easily regulated."

" The apparatus consists of a glass tube (Fig. 30), 10 millimeters in diameter, 70 millimeters high,

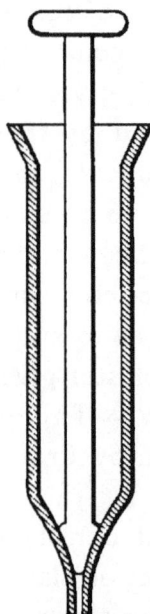

FIG. 30.

widened into a funnel at top and terminated at bottom in a point, with an orifice 1 millimeter in diameter. This orifice is closed on the interior a little above the bottom, by a small emery stopper, fastened to a long glass rod, which issues from the top of the tube.

" To make an experiment, the stopper is wet with water (grease is dissolved by the liquid), to prevent leakage. Two grammes of cement are introduced, then five cubic centimeters of liquid (density 2.95). It is then agitated in a lively manner with a small platinum thread bent around into a hook, in order to drive out the bubbles of air and thoroughly mix the cement in suspension in the liquid. Finally, it is allowed to settle for an hour; after that time there is formed two layers—the cement at the bottom and the slag at the top. The stopper of emery is lightly raised by the rod to which it is attached, letting out the cement and part of the liquid, and is then replaced. The cement is caught upon a filter, through which the liquid passes into a flask, when it is ready for another operation.

" The slag and remainder of the liquid is received upon another filter. Finally, the tube and filters are washed with benzine and dried; the cement and slag are weighed, and analyzed chemically if thought proper."

The cost is said to be the principal objection to this method, although but a small quantity, a part of a cubic centimeter of iodide of methylene, is required for each operation.

ART. 65   ABRASIVE TESTS.

Cement to be used for sidewalks, floors, or artificial stone is sometimes submitted to tests for resistance to abrasion. This test is frequently employed in Germany, the apparatus designed by Prof. Bauschinger being used. This apparatus consists of a cast-iron disk 122 centimeters in diameter and 3 centimeters thick, mounted to rotate horizontally at about 20 revolutions per minute. Specimens 6 × 6, 10 × 10, or 12 × 12 centimeters in section are employed.

They are held upon the disk with a pressure of 30 kilogrammes, and 20 grammes of standard sand are

FIG. 31.

added for each 10 turns. 200 turns are given, and the loss in weight of the specimen is determined.

Fig. 31 shows a similar apparatus made by the

Riehle Bros. Testing-machine Co., and used for brick and stone tests. Abrasion tests, when employed, are usually made for both neat cement and sand-mortar. The test of mortar as used in practice is evidently the more important. Resistance to abrasion varies with the character of sand used, and for sand-mortar depends upon the adhesion of the cement to the sand and the hardness of the grains of sand. Sand-mortars with moderate proportions of sand give better resistance to abrasion than neat cement.

ART. 66. TESTS FOR POROSITY.

Tests for the porosity of mortar are interesting in studying the properties of the materials and methods of gauging them. The porosity may often be a matter of importance as affecting the durability of mortar subjected to the action of disintegrating agencies. These tests are not employed for the reception of material. The porosity depends to a much greater extent upon the quantity of water used in gauging, and the degree of compression used in forming the briquette, than upon the character of the cement or sand employed, although probably the fineness of these materials has some influence.

The test for porosity consists in determining the ratio of the volume of voids to the total volume of the mortar. The difficulty of determining when the voids are completely filled with liquid, or when the mortar is quite dry, makes the process a somewhat uncertain one, and requires that a definite procedure be followed in order to arrive at concordant and comparable results.

The method usually followed is to measure the total apparent volume and the volume of material: the volume of voids is then the difference of the two; this divided by total volume is the percentage of porosity. To measure the total apparent volume, the simplest method is to make the block of such form that it may be directly measured. When this method is not employed the total volume may be obtained by weighing the block in a saturated condition in water and in the air; the difference between the two weights is the weight of water displaced, from which the volume may be found. In order to obtain the same state of saturation the weight in water should be taken immediately before that in air. Grease is also sometimes applied to the surface of the block to prevent change during the weighing.

To obtain the volume of solid material in the block, the difference between the weight of the block when dry, in air, and of the saturated block in water is obtained. This difference is the weight of water displaced by solid material. To secure good results the entire dryness in the first instance and the complete saturation in the second is essential. The block may be placed in warm dry air for a period sufficient to permit the weight to become constant. For this purpose it is necessary that the temperature be the same in all cases, as the amount of hygrometric water given off depends upon the temperature of drying; $100°$ to $110°$ Fahr. has been sometimes employed, and is recommended by the French Commission on Standard Tests. After the dry weight is obtained, considerable difficulty may be experienced in getting complete saturation. If the block be simply immersed, air will be retained in the voids, and a long period required to obtain a constant weight. Boiling is sometimes resorted to, but has the disadvantage of perhaps caus-ing change of volume in the mortar. The best method of expediting the test is to exhaust the air by placing the specimen in water under the receiver of an air-pump.

Prof. Tetmajer recommends * that a temperature of 110° C. be used in drying, and that the volume of solid matter be obtained by weighing dry in air and satu-rated in paraffine, and determining the volume of paraffine displaced. The block is then put in water, and the total volume obtained by measuring directly the volume of water displaced. For this purpose an apparatus is employed which is very similar to the Schuman volumenometer, Fig. 5, but with a remov-able cover to the dish to admit the specimen.

ART. 67. TESTS FOR PERMEABILITY.

The permeability of mortar is quite distinct from its porosity, and the more porous is not necessarily the more permeable. Tests for permeability, like those for porosity, are interesting and important in the studies of the properties of mortar, but are not suitable for use in specifications for the reception of material. The common test for permeability is made by forcing water through a cake of mortar under pressure. This

---

* Methoden und Resultate der Prüfung der Hydraulischen Bindemittel (Zurich, 1893).

may be accomplished either by subjecting the mortar directly to the pressure of a considerable head of water, or by subjecting the block of mortar to a small pressure from a column of water above, while the air is exhausted, forming a partial vacuum below. The latter method has been usually preferred in Europe, although the former has been made standard in France.

The apparatus shown in Fig. 32 has been frequently employed in these tests. It consists of a heavy cylindrical glass jar (*a*) with a ground upper edge, upon which is accurately fitted the second cylinder (*d*). The specimen (*c*), having a section of 20 square centimeters and a thickness of 1 centimeter, is fastened and made water-tight in the cylinder (*b*) by means of a rubber packing-ring. A ground-glass stopper covers the cylinder (*b*) and carries the graduated tube (*e*), which has a capacity of 200 cubic centimeters and is graduated to 1/2 centimeter. The space under the specimen is connected with an air-pump and a mercury manometer, a stop-cock being placed in the tube connecting with the air-pump. In using the apparatus, after the specimen has been placed in the cylinder and the cover clamped down, the air is exhausted from *a*, the stop-cock closed, and the graduated tube. filled with water to the zero mark. The quantity of water percolating through the specimen may then be read from the scale for any desired unit of time.

The arrangement shown in Fig. 33 is also used to some extent in Europe. It consists of a hollow cylindrical block, 110 millimeters in diameter and 200 millimeters high, of the mortar to be tested, into the top of which a glass tube is set with neat cement. In

making the test a rubber tube is connected with the
glass tube and with a vessel placed at such an eleva-

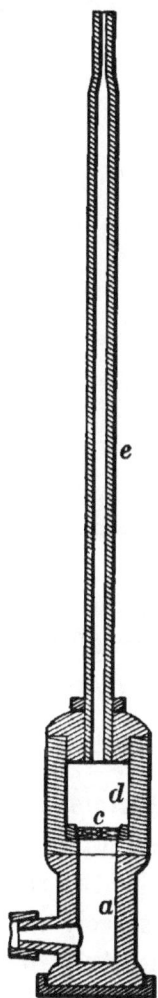

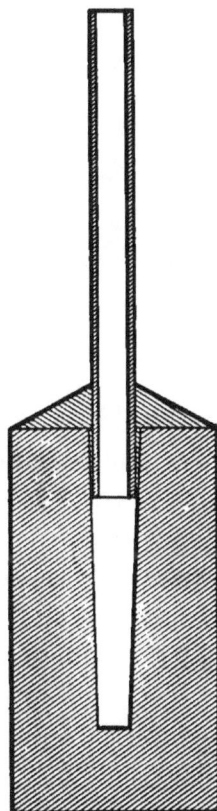

FIG. 32.   FIG. 33

tion as will give the pressure desired, which varies
according to the mortar to be tested.

A cubical block, arranged as shown in Fig. 34, is

also frequently employed.   For this purpose the
standard block as used for compressive tests may be
employed.   This form is recommended by the French

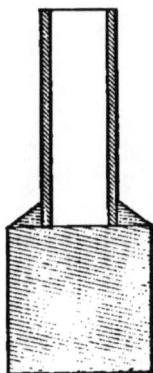

FIG. 34.

Commission upon Methods of Testing Materials, as
follows:

" The permeability of cements and mortars may be
expressed by the number of liters of water which
traverse a cube with faces of 50 square centimeters
during a given time.

" The water is supplied by a glass tube 35 millimeters in diameter and 110 millimeters high, sealed
vertically by the aid of neat cement upon the upper
face of the block.   The upper end of the tube is connected by a rubber tube with a reservoir raised to an
elevation corresponding to the head desired.   The
heads adopted, according to the permeability of the
mortar, are 0.10 meter, 1 meter, and 10 meters.

" Before the test is made, the blocks should be immersed for 48 hours, with such precautions as are
necessary to secure as complete saturation as possible.

After beginning the experiment the block is kept completely immersed.''

'' The rate of filtration is determined after 24 hours, 7 days, 28 days, 3 months.''

'' The determinations are made for three blocks, the mean results being taken for the two blocks most concordant.''

'' The *standard test* for permeability is to be made upon standard plastic mortar 28 days old, kept under water.''

'' For tests upon mortars of different ages and compositions, it is recommended to employ 1 to 2 and 1 to 5 mortars made plastic, and 7 days, 28 days, and 3 months old.''

The permeability of mortars to gases is a matter upon which experiments are very desirable, but concerning which comparatively little is known.

ART. 68.   FROST TESTS.

Tests are frequently made upon cement-mortars to determine the effect of freezing before setting or while the mortar is still comparatively fresh, for the purpose of investigating the safety of using the mortar in freezing weather, or the best method of so using it.

These tests are usually made by exposing briquettes to freezing temperatures, either by taking advantage of natural low temperatures, or by using a freezing mixture, and comparing the results of tensile tests upon briquettes so treated with those kept under normal conditions.

Tests of this kind may be of much value in showing

the relative properties of various materials, and often give very interesting results. They are to be used with caution, however, in determining from them the probable effect of freezing upon work in which the mortar may be used. Frequently the results of briquette tests, and the action of the material in large masses in construction are not concordant. Injury to work may perhaps result not only from the injurious effect of frost upon the strength of the material, but also from expansive action upon the mass of mortar, after setting, while still too weak to offer effectual resistance to distortion.

ART. 69.   TEST FOR YIELD OF MORTAR.

In some laboratories it is the custom to make tests of the yield of mortar obtained from given weights of the materials employed. This may sometimes be of importance as affecting the economy of use of various materials, while a study of the differences obtained with different material is interesting.

In conducting such tests it is evidently necessary to adopt a standard method of gauging the ingredients, the volume of the resulting product being much affected by variations in manipulation. The method employed is usually to measure directly the volume of paste obtained by mixing to standard consistency a unit weight of neat cement or of cement and sand in proper proportions. For this purpose the paste is put in a graduated glass cylinder, care being taken to eliminate all of the air-bubbles.

Sometimes the test is made by making blocks of the

paste, which are allowed to set, their volumes being subsequently obtained by greasing their surfaces and taking the difference of weight in air and water.

ART. 70. TESTS OF SAND.

Tests of the sand to be used in mortar are of much value in determining the relative value of different sands, as well as in studying the effect of variations in the nature, form, or size of grains of which it may be composed.

Complete tests should include an examination of the nature of the sand, its fineness as shown by the amount retained by various sieves; the form of grain, which may be examined under a glass; its specific gravity or the weight of a unit volume; its mineralogical character. Tests of the tensile strength of mortar made from the sand to be tested as compared with similar tests made upon standard sand are of most importance as indicating the value of the sand to combine with cement in forming mortar.

CHAPTER IX.

CEMENT-MORTAR AND CONCRETE.

ART. 71. SAND FOR MORTAR.

As hydraulic cement is commonly mixed with certain proportions of sand, when used in construction, the nature and quantity of sand used, and the method of manipulating the materials in forming the mortar, have nearly as important an effect upon the final strength of the work as the quality of cement itself.

In testing cement, a standard sand is usually employed. This sand may be obtained quite uniform in quality. In the execution of work, however, local sand must generally be employed; this varies widely in character, and should always be carefully considered upon any important work, where the development of strength and lasting qualities in the mortar is of importance.

The importance of the quality of sand in mortar is not commonly appreciated, and but little attention is usually given to securing good sand even when the cement is subjected to rigid requirements.

Mr. Newman, in his book on concrete, in speaking of the materials used in concrete, says: " Considering the very varied character of sand and gravel, it seems that more attention should be given to the particularization of the sand and gravel, remembering the locality of the work in each case, and the geological features of the district from which, for reasons of economy, the sand or gravel must be obtained.

" The value of it from an engineering point of view may be very different, even in a small area; and to be most particular as to the character and quality of Portland cement, and apparently regardless of that of the sand and gravel, although the latter form 85% to 93% of the volume of concrete at the time of mixing, is hardly capable of vindication, especially as Portland-cement concrete should be a monolithic mass, and the effect of sand is to retard induration and decrease strength."

The *chemical nature* of the sand does not appear to have any important bearing upon its usefulness in mortar. Silicious sand is sometimes thought to exercise a slow puzzolanic action, and perhaps aid somewhat in the final hardening of the mortar. It is usually the best sand for the purpose. Calcareous sands are good, if not friable or composed of soft particles. Argillaceous sand is usually less desirable, and has been found in some instances apparently to cause ultimate disintegration in sea-water,* although a small admixture of clay in the sand may not be objectionable, and has been shown in some instances not to decrease the strength when present to an extent not exceeding 10% of the sand.

A sand for use in mortar should be clean, and as free from loam, mud, or organic matter as possible.

---

* Annales des Ponts et Chaussées, 1890, vol. II. p. 277.

In general, the presence of any foreign matter is to be avoided.

The sand should also be as sharp as possible; if it be composed of angular grains, it will compact much closer and make a stronger mortar when used with the same proportion of cement than if it be composed of rounded grains.

Coarse sand is usually preferable to that which is very fine, provided it be fine enough to give a smooth mortar, as it gives better strength. The coarse sand presents less surface to be coated with cement, and the interstices are more easily filled. Fine sand requires more water in mixing in order to arrive at the same consistency, and thus gives usually a more porous mortar. Fine sand may, however, be desirable when an impervious mortar is the object.

The use of a mixture of grains of various sizes is usually desirable, as giving less voids to be filled by the cement; and it is frequently found that when the cement is not in considerable excess the strength obtained by such a mixture is much greater than is given by either the large or small grains alone.

This is doubtless due to the voids in the sand being more completely filled by the cement. Large grains of uniform size seem desirable where a meagre mortar is to be employed, the quantity of cement being insufficient to fill the voids and only used to coat the grains and cement them together.

Fine sand is objectionable in mortar exposed to the action of sea-water on account of the increased porosity.

In using quick-setting cement the dryness of the

sand is a matter of importance; if the sand be damp, when the mixture of sand and cement is made, sufficient moisture may be given off to induce hydration previous to the addition of the water. With slow cements this is of less consequence.

M. Candlot found * cement left in contact with sand which was slightly damp, not sufficiently so to cause the cement to set, was greatly modified in action, probably through the hydration of the aluminate of lime.

Cement left ten minutes in contact with sand containing 3% of moisture, and then sifted out, had its time of setting increased from a few minutes to several hours. When the sand was very wet, the action was more serious and a loss of strength resulted.

ART. 72.   PROPORTIONING MORTAR.

The relative proportions of the ingredients to be used in mortar are usually stated as a ratio of parts of cement to sand, and the quantities are determined either by weight or volume. Cement should always be measured by weight on account of the variation in volume caused in packing, and the difficulty of measuring by volume always in the same way, while the sand may conveniently be measured by volume.

The proportions of sand and cement to be used in any instance depends upon the nature of the work and the necessity for the development of strength or imperviousness in the mortar. The volume of in-

---

* Ciment et Chaux hydrauliques (Paris, 1891).

terstices to be filled varies with the forms and sizes of
the grains of sand, and the quantity of cement necessary to reach the same strength with different sands
varies considerably. The volume of a given weight
of sand is also greater when damp than when dry, and
the same proportion to a given volume of sand gives
a richer mortar when the sand is measured in a damp
condition than when measured dry.

The proportions most commonly used in ordinary
work are, for natural cements, one part cement to one
or sometimes two parts sand, and for Portland cement
one part cement to three parts sand. If the proportions for the mixture were regulated by the value of
. the sand the interests of economy might frequently
require changes in proportions, and would usually
demand the use of the best sand obtainable.

Good sand in a 1 to 3 mixture frequently gives
greater strength than a poorer one mixed 1 to 2, and
either mortar may give equally good results in practice.

The cement in mortar must, for the best results,
both coat the grains of sand so as to cause them to
adhere to each other, and fill the voids between them.
Mortar to be exposed to the action of water, particularly sea-water, should always contain a surplus of
cement over what is necessary to fill the voids in the
sand.

M. Candlot gives as a minimum for mortar to be
used in sea-water a proportion of 600 kilogrammes of
cement to a cubic metre of sand, which is to be increased when the cement is not finely ground.

Fine sand is to be avoided, if possible; when used,
the proportion of cement to be increased.

The complete filling of the voids in the sand so as to exclude the water from the interior of the mass and prevent the action of the magnesian salts upon the cement is in such work a matter of first importance.

When the mortar is to be subjected to the action of fresh water its permeability may not be a matter of so great consequence, and in many instances less rich mortar may frequently be used to advantage, provided sufficient strength be obtained for the given purpose. The carbonizing of the lime forms a protection to the mass of mortar, which is not subject, as in sea-water, to the action of the magnesian salts.

ART. 73. GAUGING MORTAR.

In mixing cement-mortar, the cement and sand are first thoroughly mixed dry, the water then added, and the whole worked to a uniformly plastic condition. The value of the mortar will depend upon the thoroughness of the operation; the cement must be uniformly distributed through the sand during the dry mixing, while thoroughly working the mass after the addition of the water will greatly increase its strength. In mixing by hand, by the ordinary method, a platform or box is used; the sand and cement are placed upon the platform in layers, with a layer of sand at bottom, and then turned and mixed with shovels until properly distributed through the mass. The material is then formed into a ring, or a mound with a crater at the centre, and all the water necessary added at once, after which the material is thrown up from the sides until the water is all taken up, and then worked into a plastic condition.

In order to secure proper manipulation of the materials on the part of the workmen, it is quite common to require that the whole mass shall be turned over a certain number of times with the shovel, both dry and wet.

The mixing should be quickly and energetically done, only such quantity being mixed at once as can be used before the initial set of the mortar takes place.

The cement should not be left in contact with the sand for any considerable time before being used, or a considerable quantity should not be mixed dry and left to stand until wanted, as the moisture commonly in the sand will to some extent act upon the cement.

Upon large works mechanical mixers are frequently employed, with the advantage of greatly lessening the labor of manipulating the material, and also of insuring thorough mixing.

The quantity of water to be used in gauging mortar can be determined only by experiment in each instance. It depends upon the nature of the cement and sand, and upon the proportion of sand to cement. The water may be considered as made up of two parts— that necessary to gauge the neat cement to a paste, and that required to wet the surfaces of the sand. The first varies directly with the quantity of cement, the second with that of the sand.   Fine sand requires more water than coarse sand to reach the same consistency, and mortar of fine sand should be made a little more wet than when of coarse sand, to give the best results in practice. The quantity of water also varies with the dryness of the sand and its porosity.

The amount of water to be used in mixing mortar for ordinary masonry is such that the mortar when properly mixed shall have a stiff plastic condition. It should not be a soft, semi-fluid mass. The proper consistency is described by M. Candlot as such that if a ball of mortar be formed in the hand and allowed to fall through a small height, it should neither lose its form nor crack; the ball should not be wet enough to stick to the hand. The best results are usually obtained by mixing with as little water as will admit of proper manipulation in the work, and wetting the surfaces with which it is to be in contact.

In all cases the proper quantity of water should first be determined by experiment, and afterward, in preparing the mortar for use in work, the required quantity should each time be added by measurement. The addition of water little by little, or from a hose, should never be allowed.

ART. 74.   PREPARATION OF CONCRETE.

Concrete is any mixture of mortar with coarse material, usually gravel or broken stone, the office of the mortar being to bind together the pieces of the aggregate and fill the spaces between them. In engineering work the mortar for concrete is commonly formed from hydraulic cement, the term *beton* being also frequently used to designate hydraulic concrete.

In preparing concrete by hand the mortar is mixed in the usual manner; then the stone is spread over the top of the layer of mortar and thoroughly mixed with it by turning with shovels. The stone should be sprinkled before being mixed with the mortar, sufficiently to wet its surfaces and prevent the absorption of the water from the mortar, thus promoting the adherence of the mortar to the aggregate.

Mortar for concrete should never, as is frequently done, be reduced to a fluid state; not only is the resulting strength of the mortar reduced by so doing, but it cannot be properly mixed with the aggregate to form a homogeneous mass, as the cement washes out of the mixture. The mortar should, however, be sufficiently soft to mix readily with the aggregate to a cohesive mass, which may be placed and compacted without difficulty in the work. The consistency of concrete must in each instance depend upon its nature and method of use; the greatest strength may usually be attained by mixing somewhat dry and heavily ramming, the mortar being of such consistency that the concrete becomes somewhat jelly-like, water being brought to the surface in ramming. An extreme either of dryness or wetness may be injurious.

Mechanical mixers are frequently employed for preparing concrete, and are very useful in saving labor where considerable quantities are used. There are a number of forms which have proven effective in use, but it seems unnecessary to enter into a discussion of them here.

The aggregate used for concrete should be as hard and durable as possible, and that of angular form is preferable to rounded. Angular forms give a greater surface for the adherence of the mortar in proportion to volume while leaving a less volume of interstices to be filled with mortar. The materials should be uniform in quality. Where gravel is used which varies in quality, it should be blended by mixing in order to obtain uniform strength in the concrete. Porous aggregates are to be avoided, as they are likely to absorb the cement. When the materials are absorbent, they should be saturated in sprinkling before using, in order to avoid withdrawing water from the mortar before setting takes place.

The quantity of sand used in concrete should be such as is necessary to fill the voids in the aggregate, while the quantity of cement depends upon the strength necessary in the work under consideration. When the concrete is required to be water-tight, the amount of cement paste must be sufficient to fill the interstices in the mixture of stone and sand. The quantity of sand necessary to fill the interstices in the stone may be determined by filling a measure with stone as closely as possible, and then measuring the quantity of water which can be poured into the measure; this gives the volume of sand required. If the proper quantity of damp sand be added to the stone in the measure by shaking it down so as to fill the voids, the volume of water which can then be put in the measure is the volume of cement paste necessary to fill the voids in the aggregate.

The strength of concrete usually varies nearly in proportion to the amount of cement used in forming it. When a strong concrete is desired, it should be obtained by increasing the richness of the mortar in cement, not by increasing the proportion of mortar to large material above the point at which the sand fills the interstices in that material. If the proportion of sand be less than this, the resulting concrete will be porous and not thoroughly solidified; if it be greater, the excess of sand may be an element of weakness in the concrete.

In the use of concrete in considerable masses the main body of the work is sometimes formed of very weak concrete, with a facing of stronger water-tight concrete to protect it. This weak concrete is frequently formed by omitting the sand altogether and simply coating the stone lightly with neat cement, causing the stones to adhere to each other, thus forming a mass sufficiently firm for foundations in many locations when protected by a covering of richer concrete. The voids in a mass of ordinary broken stone vary from about 4/10 to 5/10 of the volume, depending upon its uniformity in size. Where there is considerable variation in size, the voids may be somewhat less. When the interstices are to be filled, it is desirable that the aggregate contain material of various sizes, to reduce the volume of interstices. For this reason small gravel is sometimes mixed with broken stone in the preparation of concrete.

The proportions in common use for concrete of Portland cement vary from 1 part cement, 2 parts sand, and 5 parts broken stone to 1 part cement, 4 parts sand, and 8 or 10 parts broken stone or gravel. Usually the mortar is made richer when natural cement is used. The proportions, of course, vary with the character of the materials to be used as well as that of the work to be done, and can only be properly determined by the exercise of good judgment, in the light of experience.

ART. 75. YIELD OF MORTAR AND CONCRETE.

The volume of mortar formed by mixing given quantities of cement and sand depends upon the densities of the materials and the volume of interstices in the sand.   It is affected also by the method of preparing the mortar, the uniformity of the mixing, and the degree of compactness given.

The net volume of materials entering into the composition of mortar or concrete is readily found from their weights and densities, but it represents only approximately the resulting volume.   An accurate knowledge of the yield of any particular mixture is to be obtained only by experiment upon the materials to be employed.

Portland cement is usually sold in barrels containing about 375 lbs.   Natural cements are lighter, and are put up in barrels of 260 to 320 lbs.   Barrels of Rosendale cements usually contain 300 lbs.

The amount of neat cement paste made by a given weight of cement powder varies with the specific gravity of the cement and the amount of water necessary in gauging.   The lighter cements require more water and yield less paste for a given volume of cement than the heavier ones.   To form a cubic foot of plastic paste requires usually from 75 to 90 lbs. of natural-cement powder; about 80 to 85 lbs. of Rosendale cement being required, while about 95 to 100 lbs. of Portland cement are necessary.

In mixing sand-mortar, where the cement and sand
are proportioned by volume measured loose, the
quantities required to form a cubic yard of mortar are
approximately as follows:

	Natural Cement. Pounds.	Portland Cement. Pounds.	Sand. Cu. Yd.
1 to 1 mortar....	1050 to 1250	1350 to 1530	.65 to .70
1 to 2 "	640 to 720	810 to 920	.80 to .85
1 to 3 "	500 to 575	620 to 690	.93 to .96
1 to 4 "	400 to 460	500 to 575	1.00
1 to 5 "	320 to 375	400 to 460	1.00

For concrete, when the aggregate is broken stone
of uniform size, it is necessary, in order to fill the interstices with mortar, that the volume of mortar be 50%
to 60% that of the aggregate. For such concrete a
mixture of about .9 cubic yard of broken stone with
.50 to .55 cubic yard of the mortar as given above
yields about one cubic yard of concrete. This gives
the proportions sometimes employed for strong concrete: 1 part cement, 2 parts sand, and 4 parts broken
stone; or 1 part cement, 3 parts sand, and 5 parts
broken stone.

Where the stone is more irregular in size, or if
gravel of smaller size be added, a smaller proportion
of mortar may give good results. Thus, .9 cubic yard
of broken stone with .4 cubic yard of gravel and .3
cubic yard of mortar has been found to yield 1 cubic
yard of good concrete. This, using 1 to 2 mortar,
gives the proportion 1 part cement, 2 parts sand, 3
parts gravel, and 7 parts broken stone.

When the amount of mortar used is less than that
indicated above, and the interstices in the aggregate

are not filled, the yield of concrete is about equal to the volume of aggregate employed.

ART. 76. MIXTURES OF LIME AND CEMENT.

Slaked lime is sometimes mixed with hydraulic cement for the purpose of decreasing the cost of construction. Experiments seem to indicate that a very considerable percentage of lime may frequently be added without material loss of strength in the mortar.

With Portland cement the addition of lime weakens the mortar somewhat, the decrease in strength augmenting rapidly as the proportion of lime increases.

With some American natural cements it has been found that a certain amount (sometimes 30% to 40%) of lime may be added without sensibly decreasing the strength of the mortar or impairing its hydraulic properties. This may be due to puzzolanic action on the part of the cement, which is of high hydraulic index.

When mortar is not to be used under water, and only moderate strength is necessary, it may often be economical to form the mortar by the admixture of lime, better results being obtained than by using a higher proportion of sand. The mortar thus formed is less porous than that made with a larger proportion of sand, and it is also more plastic and easier to work. In making the mixture the lime is ordinarily slaked in the usual manner and used in the form of paste, although it may be slaked to powder and mixed dry with the cement. By the first method the thorough slaking of the lime is insured.

The admixture of lime causes the cement to become

slower-setting, the quick-setting cement being affected more strongly than the less active ones.

In France a small proportion of Portland cement is sometimes added to hydraulic lime for the purpose of accelerating the setting and increasing the strength of the lime.

Mixtures of natural and Portland cement have frequently been used in the United States for the purpose of modifying the action of the quick-setting material or of cheapening construction. They seem generally to give results compounded of those which would be obtained by using them singly.

The results of all these mixtures will be found to vary with the particular cement employed, and the effect can only be known by trial in each instance. In all cases, to get good results, the mixtures must be very intimate.

ART. 77. THE FREEZING OF MORTAR.

Mortar of good Portland, or of many kinds of natural cement, is not injured by freezing, when frozen before it is set. The cement sets with extreme slowness, if at all, while frozen, but after thawing it sets and hardens properly. Mortar frozen for short periods—a few days—does not set while frozen, but the experiments of Mr. Cecil B. Smith at McGill University seem to show that if kept frozen for a sufficient period it may finally set while frozen.

The hardening of cement which has been frozen is much more slow than that unfrozen, but it may ultimately gain the same strength.

Masonry constructed during freezing weather is frequently injured by freezing, notwithstanding the fact that the cement itself shows no loss of strength due to freezing. The effect of frost coming upon the work before it is fully hardened is frequently to distort or cause unequal settlement in it, and sometimes repeated freezing and thawing gradually causes the mortar to be thrown out of place or perhaps to become cracked and disintegrated on the outside. The construction of cement masonry during freezing weather is therefore usually more or less hazardous, unless some means be adopted of preventing the freezing action. Many instances may, however, be cited where extreme cold has not injured work constructed, without such precaution, with Portland-cement mortar, and it is claimed by many engineers that Portland may be used with impunity in freezing weather, but usually it is not placed in work while a freezing temperature prevails. It is commonly agreed that most natural cements should not be used when a very low temperature is likely to reach the work in advance of it having attained good strength, and instances are numerous of work having been injured by changing temperature of winter weather, although it may not have frozen for considerable time after setting.

Salt is quite commonly used in cold weather to prevent the freezing of mortar while it is soft. A strong solution, frequently a saturated one, is employed. The salt, by preventing the freezing of the water, prevents any distorting or disrupting action upon the work due to the change in volume of the mortar. The use of salt considerably decreases the activity of the cement, and mortar may stand in a soft condition at freezing temperatures and finally set when the temperature becomes sufficient to induce action.

The loss in early strength of cement-mortar which has been mixed with salt water on exposure to low temperature before setting is usually greater than that of mortar without the salt and exposed at the same temperature.

The effect of salt upon the strength of various kinds of cement is quite different. In nearly all, the strength of mortar kept in air is increased by its use. When the mortar is kept under water, most cements have an access of early strength from the use of salt, which is lost later, the final strength being somewhat reduced. This is true of most Portland cements. Some natural cements suffer a material loss of strength when mixed with salt water, while others are entirely ruined by a low temperature with or without the use of salt. Care should always be taken to determine the action of salt and cold upon the particular cement before using it in this manner.

It is advisable in using salt to protect the mortar from contact with water immediately after setting, as sometimes salt mortar which has been exposed to low temperature may lose its cohesion if submerged soon after setting.

Soda has sometimes been employed to prevent the freezing of mortar, but its use has not become extensive, and has usually proven unsatisfactory.

Hot water should not be used in mixing mortar in freezing weather. It not only decreases the strength of the mortar, but renders it more liable to injury from frost. Heating the stones or bricks in the construction of masonry in freezing weather may be beneficial, as serving to accelerate the setting and keep the mortar from freezing while soft.

The injury done to mortar by low temperatures is probably not usually due to freezing before setting, but to alternate thawing and freezing while work is still fresh, before hardening is sufficiently advanced to render the mortar capable of adequately resisting the expansive forces. The effect of frost upon mortar which has set is similar to that upon stone or brick, and is due to the increase in volume of water freezing in its pores. Its effect therefore depends both upon the porosity of the mortar and upon the strength it possesses to resist disruption. The more rapid acquisition of strength by Portland cements may give them the advantage they possess in this regard.

Prof. Le Chatelier, from his experiments upon the matter, concludes as follows: " This disintegration, like that of frozen stone, is more easily accomplished when the mortar offers small mechanical resistance, when the total volume of voids is large, and when the dimensions of each separate void are small. When the voids are sufficiently large, the ice breaks with a pressure less than that which will rupture the mortar. For this reason mortars of large sand are less affected, the voids being larger and less numerous."

ART. 78. POROSITY AND PERMEABILITY OF MORTAR.

The *porosity* of cement-mortar depends rather upon the manipulation of the materials in gauging than

upon the quality of the cement.   When the quantity
of cement is insufficient to fill the voids in the sand,
spaces are left which permit the absorption of water
without increasing the volume of the mortar.

In gauging mortar air-bubbles attach themselves to
the wet sand, the number of which is greater as the
mortar is mixed more wet.   Working the mortar tends
to eliminate them.   Voids in the mortar are also
caused by the evaporation of surplus water used in
mixing.   Porosity is greater as the quantity of water
used in gauging is increased and as the sand used is
finer.

The *permeability* of cement-mortars varies with the
quality of the cement and the circumstances of its use.
Mortar of neat Portland cement may be made practically impermeable under a considerable head of
water; that composed of cement and sand seems
always more or less permeable, but when properly
proportioned and mixed eventually permits very little
water to pass.

The permeability of mortar decreases rapidly with
its age: for the first few days or weeks after mixing,
water passes quite freely through it, but as the hardening process approaches completion its power of
resistance is in this particular greatly augmented.

If blocks of mortar be submitted to the continuous
filtration of water, the permeability diminishes very
rapidly, and after a few months all mortars, except
those of very coarse sand and feeble proportion of
cement, become practically impermeable.

Both the porosity and permeability are less for
mortar rich in cement than for that in which the pro-

portion of cement is small. Mortar mixed dry is penetrated more readily than that mixed to a plastic or semi-wet condition. With the lapse of time, however, the mortar mixed dry, if constantly exposed to water, approaches the others in resistance to permeation. The thoroughness of mixing and degree of compacting employed are more important factors than the absolute quantity of water used in mixing

Fine sand, according to the experiments of M. Alexandre, renders the mortar more porous and less permeable than coarse sand. When the sand is of varying sizes both the porosity and permeability may be low. In any case, to attain a reasonable resistance to penetration, it is necessary that the interstices in the sand be entirely filled with cement. Cleanliness of the sand, its freedom from all foreign material, is of first importance in the preparation of impermeable mortar.

Masonry of ordinary brick or stone can only be made impervious by the application of a coating of some kind to its face. A plastering of neat cement or rich mortar may sometimes be used for this purpose, and coatings of asphalt or coal-tar have sometimes been successfully employed.

In concrete work where imperviousness is essential, it may be advisable, as with masonry, to coat the face of the concrete. In order that concrete may be reasonably water-tight, it is necessary that the quantity of cement-mortar used in preparing it be sufficient to fill the voids in the large material employed, as well as that the voids in the sand be completely filled with cement paste in making the mortar.

Art. 79.   Expansion and Contraction of Mortar.

In the use of large masses of masonry or concrete the change that is liable to occur in the volume of mortar may frequently become of importance, and it may be necessary to make provision by which changes in dimension an take place without in·ury to the work.

The coefficient of expansion for neat cement under the action of heat is, as previously stated, about the same as for iron, although it may vary in individual instances. For mortars containing sand, the coefficient is less than for neat cement.

Cements differ considerably in their behavior during the continuance of the hardening process, as to the change that takes place in the volume of the mortar. Unsound cement is apt to swell and become distorted at the commencement of the process of disintegration, and of course any considerable change of this nature indicates the probable destruction of the mortar. Perfectly sound cement, although not altered in form, is usually changed somewhat in dimensions during hardening: if the mortar be conserved in dry air, a slight shrinkage takes place; if under water, the mortar swells a little.

Prof. Swain, in a series of experiments at the Massachusetts Institute of Technology for a committee of the American Society of Civil Engineers, found that for small blocks of mortar the change was the same in all directions; that for neat cements the linear contraction in air varied from 0.14% to 0.32% for the first twelve weeks after mixing, and the linear expansion in water varied from 0.04% to 0.25%. When sand was used the change was less, giving a contraction in air from 0.08% to 0.17%, and an expansion in water of from 0.00% to 0.08%.

The rapidity of the change varies somewhat with the activity of the cement; the conclusion being that a quick-setting cement changes more in volume than a slow-setting one.

Further experiment is desirable, that the action of the various classes of cement may be better understood.

ART. 80.   EFFECT OF RETEMPERING MORTAR.

Masons frequently mix water in considerable quantities, and, if the mass becomes stiffened before being used by the setting of the cement, add more water and work again to a soft or plastic condition. After the second tempering the cement is much less active than at first, and remains a longer time in a workable condition.

This practice is not usually approved by engineers, and is not permitted in good engineering construction, although there is some dispute as to its injurious effect. M. Alexandre, from an extensive series of experiments,* concludes that no injury is usually done to mortar by retempering, provided sufficient water be added to make the mortar plastic at the second work-

---

* Annales des Ponts et Chaussées, 1888, vol. I. p. 375.

ing. The hardening of mortar so treated is very slow at first, but it may subsequently (the tests extend over three years) gain as much strength as when gauged immediately upon mixing.

It is also frequently claimed that the adhesive properties of mortar are improved by giving it the " second set." The common practice of masons who set fireplace tiling and similar work is based upon this idea. Further experiment to determine this point would be interesting.

The results of experiments other than those already quoted have seemed to show that in some instances injury is done to mortar by retempering, some cements even refusing to set the second time. Until more is known of the action of the material when subjected to this treatment, it seems advisable to mix only such quantity at once as may be used before the initial set of the cement, and to reject any material that may have set before being placed in the work.

APPENDIX.

SPECIFICATIONS FOR THE RECEPTION OF CEMENT.

IN the discussion which has been given of the various tests applied to cement, the requirements of specifications have been considered, but it is thought desirable to append a few actual specifications to show the requirements employed in practice.

Some engineers who use considerable quantities of cement of a few brands, or of a single brand, employ no specifications, but depend upon the reliability of the brand, or perhaps upon occasional examinations to show that the material is up to standard. This is frequently the practice upon ordinary railroad work, and in some instances, where the use is continuous and private contracts may be made for the material, is quite satisfactory.

Some specifications (quite frequently employed) make no mention of any specific requirements, but call for cement which shall be satisfactory to the engineer in charge of the work, and lodge in him full power to accept or reject the material. The effect of such a specification depends upon the circumstances

under which it is employed. When used as a specification for work, in open competition, it is very apt
to prove unsatisfactory; but when it represents the
established practice of an office dealing with particular
brands of cement, it may not be objectionable.

There seems, generally, to be no good reason for
stating specifications in an indefinite manner. The
conditions to be imposed may as easily be plainly
stated, and thus leave no doubt as to the requirements.
In some instances, however, engineers consider a
specification of this kind advantageous where the
cement to be used is limited to a few accepted brands
(particularly with natural cements), the tests to be imposed varying with the brand and being only for the
purpose of showing that the various lots of cement are
normal in action. In such a case the tests to be imposed are usually an understood, though not an expressed, part of the specifications.

The specifications which follow, A, B, C, D, E, are
thought to represent fairly well the range of requirements in the best American practice

Specifications " A " represent about the common
requirements for ordinary masonry, but are more
complete and explicit in statement than is common.
The tensile test requirements vary considerably in
various specifications, being frequently much higher
where permanent laboratories are maintained, as in
specifications " D " and " E."

Specifications " B " have much the same requirements, but apply the Faija warm-water test for soundness.

Specifications " C " are intended for cement to be

used for heavy work in sea-water. Soundness is assured by applying the hot test, combined with a limit of sulphuric acid and magnesia. The requirements are high, but may be met without difficulty by most of the leading brands of Portland cement, and are considered necessary because of the importance of the work and its exposure to the action of sea-water.

It is quite common to place considerable reliance upon the brand in selecting cement, and in many specifications the approval of the brand is an essential condition. This, seemingly, places an arbitrary power in the hands of the engineer, but may often increase the likelihood of getting good material, while perhaps saving the trouble and loss of time incident to testing material obviously unfit for use.

A.

PENNSYLVANIA RAILROAD COMPANY.
P., W. & B. R. R.　N. C. Ry.　W. J. & S. R. R.

STANDARD SPECIFICATIONS FOR CEMENT.

Cement will be tested by the Railroad Company, upon its receipt, in accordance with the following specifications, and such tests are to be final as determining the matter of its acceptance or rejection.

The cement for testing shall be selected by taking, from each of six well-distributed barrels in each carload received, sufficient cement to make five to ten briquettes; these six portions, after being thrown together and thoroughly mixed, will be assumed to represent the average of the whole car-load.

Fineness. — Not more than 10% of any cement shall fail to pass through a No. 50 sieve (2500 meshes per square inch), wire to be No. 35, Stubb's wire-gauge.

Cracking. — Neat cement, mixed to the consistency of stiff plastic mortar, and made in the shape of flat cakes, two or three inches in diameter and one-half inch thick, with thin edges, when hard enough shall be immersed in water for at least two days. If they crack along the edges or become contorted, the cement is unfit for use.

Tensile Strength. — The test for tensile strength shall be made with briquettes of standard form, as recommended by the American Society of Civil Engineers, moulds for which will be furnished from the office of the Engineer Maintenance of Way. They must have an average tensile strength not less than that given in the table below:

	One Day.	One Week.	Four Weeks.
American Natural Cement:			
Neat.............................	70	95	150
1 sand to 1 cement..................	...	50	120
2 sand to 1 cement..................	...	30	60
American and Foreign Portland Cement:			
Neat.............................	100	300	450
2 sand to 1 cement..................	...	120	175

Proportion of Water. — The proportion of water used in making briquettes varies with the fineness, age, and other conditions of the cement and the temperature of the air, but is approximately as follows:

Neat Cement.—Pórtland, 20% to 30%. Natural, 20% to 30%.

1 part cement, 1 part sand, about 15% total weight of cement and sand.

1 part cement, 2 parts sand, about 12% total weight of cement and sand.

Mixing.—The cement and sand, in proper proportions, shall be mixed dry, and all the water specified added at one time, the mixing to be as rapid as possible to secure a thorough mixture of the materials, and the mortar, when stiff and plastic, to be firmly pressed to make it solid in the moulds without ramming, and struck off level.

Moulding.—The moulds to rest directly on glass, slate, or other non-absorbent material. As soon as hard enough, briquettes are to be taken from moulds and kept covered with a damp cloth until immersed.

In the one-day test briquettes shall remain on the slab for one hour after being removed from mould and twenty-three hours in water. In one week or more test, briquettes shall remain in air one day after being removed from moulds and balance of time in water.

Briquettes are to be broken immediately after being taken from the water. Stress to be applied at a uniform rate of four hundred pounds per minute, starting each time at zero.

No record to be taken of briquettes breaking at other than the smallest section.

Sand.—The sand used in test shall be clean, sharp, and dry, and be such as shall pass a No. 20 sieve (400 meshes per square inch), wire to be No. 28, Stubb's wire-gauge, and to be caught in a No. 30 sieve (900 meshes per square inch), wire to be No. 31, Stubb's wire-gauge.

Water.—Ordinary fresh, clean water, having a temperature between 60° and 70° Fahr., shall be used for the mixture and immersion of all samples.

Proportions.—The proportions of cement and sand and water shall in all cases be carefully determined by weight.

In preparing briquettes for test, sufficient material is to be taken to make one briquette at a time, and enough of water being added to make a stiff plastic paste as above stated.

The temperature of the testing-room not to be below 45° Fahr.

By order of the General Manager.

JOSEPH T. RICHARDS,
Engineer Maintenance of Way.

OFFICE OF THE ENGINEER OF MAINTENANCE OF WAY
PHILADELPHIA, January 10, 1897.

B.

SPECIFICATIONS OF THE CANADIAN PACIFIC RAILWAY COMPANY.

Mr. P. ALEX. PETERSON, *Chief Engineer.*

(Extract from General Specifications for Rubble Masonry.)

" 5. The cement used in making concrete or mortar shall be freshly ground Portland cement of approved brand, or such other cement as the Engineer may approve. It shall weigh not less than 110 lbs. to the struck bushel; and not less than 90% of it shall pass through a sieve containing 2500 meshes per square inch. The tensile strength of the neat cement after being kept in water at a temperature of about 60° Fahr. for seven days shall not be less than 350 lbs. to the square inch. Also, when mixed with one part of cement to three parts of sand by measure, it shall stand 170 lbs. to the square inch at the end of 28 days.

" In preparing sand for this test, sand shall be rejected which passes through a sieve made of No. 31 wire with 900 meshes to the square inch, or which will not pass through a sieve made of No. 28 wire with 400 meshes to the inch.

" A pat made and submitted to moist heat and warm water at a temperature of about 100° Fahr. shall show no sign of blowing in 24 hours.

" Cement shall be tested by the Engineer on delivery; and it shall be kept in a dry place, and in as good order as when delivered, until it is used."

C.

EXTRACTS FROM SPECIFICATIONS FOR FURNISH- ING PORTLAND CEMENT FOR WALLABOUT IMPROVEMENT, BROOKLYN, N. Y.

W. E. BELKNAP, *Engineer.*

" 9. All the cement to be furnished under this contract must be of the class of such material known as high-grade ' Portland ' cement free from lumps, dry and finely ground, and unless as otherwise specified, must be of one or more of the following brands, known as ' Dyckerhoff,' ' Alsen's White Label,' and ' Stettiner Star Brand.' Cement of other brands

may be furnished, provided the Contractor submits
proof satisfactory to the Engineer that it has been used
in making large masses of concrete, which have been
exposed to the action of sea-water for at least two
years previous to the date of this contract, and that
such concrete now shows no signs of deterioration
which might be imputed to defective qualities in the
cement.

" 10. All the cement shall be composed of lime,
silica, and alumina in their proper forms and proportions, be as free as possible from all other substances,
and contain no adulterant in injurious proportions.
The ratio of the weight of silica and alumina to the
weight of the lime in the cement shall not be less than
45/100.   The cement shall not contain more than 3%
of magnesia nor more than 1% of sulphuric acid.

" 11. The cement shall not have a lower specific
gravity than 3.10.

" 12. All the cement shall be of a fineness so that
99% by weight shall pass through a No. 50 sieve (2500
meshes per square inch) of No. 35 wire; 90% shall pass
through a No. 100 sieve (10,000 meshes per square
inch) of No. 40 wire, and 70% shall pass through a
No. 200 sieve (40,000 meshes per square inch) of No.
45 wire.   All wire numbered as per ' Stubb's ' gauge.

" 13. The cement must not take its ' initial ' set in
less than 30 minutes after mixing.   It shall take its
' hard ' set in not less than 3 hours and in not more
than 8 hours. .

" The cement will be said to have attained its
' initial ' and its ' hard ' set when it bears without
indentation respectively a wire of 1/12 inch diameter

loaded to weigh 1/4 pound, and a wire of 1/24 inch diameter loaded to weigh 1 pound, it having been previously mixed neat with about 25% of its weight of water, and worked for from one to three minutes into a stiff plastic paste.

" 14. All the cement shall be capable of developing a tensile strength under various conditions as follows:

Age.		Tensile Strength in Pounds per Square Inch.
Mixed neat with about 25% of water by weight and worked to stiff plastic paste.	24 hours, in water after hard set 7 days, 1 in air, 6 in water, 70° 28 " 1 " " 27 " " "	150 400 600
Mixed with 3 parts sand by weight, and about 12% of combined weight of sand and cement of water to stiff plastic paste.	7 days, 1 in air, 6 in water, 70° 28 " 1 " " 27 " " "	150 240

' To determine the tensile strength, 4 briquettes of the cement under each of the above conditions will be broken in a ' Riehle ' or Fairbanks or other testing-machine satisfactory to the engineer.

" The sand to be used in making briquettes will be clean, dry crushed quartz, trap-rock or granite, passing a No. 20 sieve of No. 28 wire, and caught on a No. 40 sieve of No. 31 wire ' Stubb's ' gauge. The briquettes will be of the form recommended by the American Society of Civil Engineers.

" 15. All cement must be sound in every respect,

and show no indications of distortion, change of volume, or blowing when subjected in the form of pats to exposure in air and fresh and sea water of temperature from 60° to 212°, as follows:

" The pats will be made of neat, unsifted cement, mixed with fresh water to the same consistency as before stated for briquettes, and will be about 3 inches in diameter, having a thickness at the centre of about 1/2 inch, tapering to about 1/8 inch at the edges. They will be moulded on plates of glass and kept thereon during examination.

" (a) One or more of these pats will, when set ' hard,' be placed in fresh water of temperature between 60° and 70° for from 1 to 28 days.

" (b) One or more of these pats will be allowed to set in moist air at a temperature of about 200° for about 3 hours. It will then be placed and kept in boiling water for a period of from 6 to 24 hours.

" (c) One or more of these pats will be allowed to set in moist air at a temperature of about 100° for 3 hours; it will then be placed and kept in water of temperature of 110° to 115° for a period of from 24 to 48 hours.

" (d) One or more of these pats may be subjected to any or all of the above indicated tests ((a), (b), and (c)), using sea-water instead of fresh water.

" (e) One pat will be kept in the air for 28 days and its color observed, which shall be uniform throughout, of a bluish gray, and free from yellow blotches.

" A failure to pass test (b) will not necessarily cause the rejection of the cement, provided it passes the other tests for soundness as noted in (a), (c), (d), and (*e*), and is satisfactory in other respects to the engineers.

" 16. All the above tests may be modified and other tests, in addition thereto or in substitution therefor, be required, at the discretion of the engineer, to practically determine the fitness of the cement for its intended use.

" 17. It is agreed by the party of the second part that he will pay all the costs of testing the cement to determine its composition, quality, and character to the satisfaction of the engineer.

" 18. It is further agreed that the tests shall be made in the manner as indicated on the schedule on file in the office of the Engineer, and that they shall be made by the Engineer or by one or more of the following parties: R. W. Hildreth & Co.; Riehle Bros. Testing-machine Co.; E. & T. Fairbanks & Co., of New York City; and Booth, Garrett & Blair, of Philadelphia, Pa.

" 19. It is agreed that as many tests shall be made to determine the composition and specific gravity of the cement as indicated in Articles 9 and 10, as shall be desired by the Engineer. It is agreed that one sample shall be tested for each 100 barrels or more to determine the quality of the cement, as per Articles 11, 12, 13, and 14.

" 20. All the cement must be furnished in the ' original package ' in strong substantial barrels, which shall be plainly marked with the brand or mark of the maker of the cement.

" 21. Each barrel must be properly lined with paper or other material so as to effectually protect the cement from dampness. Any cement oamaged by water to such an extent that the damage can be ascertained from the outside will be rejected *in toto*, and the barrels unopened. Barrels containing a large proportion of lumps will also be rejected. Broken barrels of cement, if otherwise satisfactory, will be counted as half-barrels."

" 25. It is agreed that the quality and character of any lot of cement shall be determined by the Engineer by the tests as above called for, and as per provisions of Article 30 hereof, said tests to be made upon such proportions of the whole amount of cement in any lot as he may deem proper, and it is further agreed that his decision as to the acceptance or rejection of the cement under this contract shall be final and conclusive."

" 30. It is further agreed that the Commissioner may reject and refuse to accept any cement which in his opinion is unfit for the work for which it is intended without making tests of the same and without giving any reasons for such opinion to the Contractor, but all cement must before acceptance pass satisfactorily, to the Engineer, all the tests herein prescribed."

(With reference to Article 30, Mr. Belknap says: " I might explain, in view of its seemingly giving an arbitrary power to the Commissioner, that it was inserted in order that we might avoid the trouble and delay of testing any brands which on the face of it appeared to us to be entirely unfit for the work.").

D.

SPECIFICATIONS FOR MUNICIPAL WORK IN ST. LOUIS, MO.

Mr. M. L. HOLMAN, *Water Commissioner.*

(Extract from "Specifications for Foundation of Stand-pipe No. 3, St. Louis Water Works.")

" 26. All cement for the work herein specified shall be of the best quality of American Portland. Cement without the manufacturer's brand will be rejected without test.

" 27. All cement furnished will be subject to inspection and rigorous tests, of such character as the Water Commissioner shall determine, and any cement which, in the opinion of the Water Commissioner, is unsuitable for the work herein specified will be rejected.

" 28. If a sample of the cement shows by chemical analysis more than 2% of magnesia (MgO), or more than 2% anhydrous sulphuric acid (SO$_3$), the shipment will be rejected.

" 29. To secure uniformity in cement of approved brands, all cement received on the work shall be subject to tests for checking or cracking, and to the following tests for fineness and tensile strength.

" 30. All cement shall be fine-ground, and 85% shall readily pass a sieve having 10,000 meshes to the square inch.

" 31. All cement shall be capable of withstanding a tensile stress of 400 lbs. per square inch of section, when mixed neat, made into briquettes, and exposed 24 hours in air and 6 days under water.

" 32. All cement shall be put up in well-made barrels, and all short-weight or damaged barrels will be rejected. Samples for testing shall be furnished at such times and in such manner as may be required. On all barrels of rejected cement inspection marks will be placed, and the Contractor shall in no case allow these barrels to be used.

" 33. In measuring cement for mortar or concrete, the standard volume of a barrel of cement shall be determined by comparing its net weight with the weight of one cubic foot of thoroughly compacted neat cement.

" 34. All cement for use on the works shall be kept under cover, thoroughly protected from moisture, raised from the ground, by blocking or otherwise, and dry until used. The Contractor shall keep in storage a quantity of accepted cement sufficient to secure the uninterrupted progress of the work.

" 35. Accepted cement may be re-inspected at any time, and if found to be damaged or of improper quality will be rejected. All rejected cement shall be at once removed from the work."

(Mr. Holman states that in his future specifications he intends to reduce the allowable sulphuric acid to $1\frac{1}{2}\%$.

E.

SPECIFICATIONS FOR MUNICIPAL WORK IN PHILADELPHIA.

GEORGE S. WEBSTER,
CHIEF ENGINEER.
RICHARD L. HUMPHREY.
INSPECTOR OF CEMENTS.

DEPARTMENT OF PUBLIC WORKS, BUREAU OF SURVEYS.

SPECIFICATIONS FOR CEMENT AND MORTAR.

CEMENTS

1. **Inspection.**—All cements shall be inspected, and those rejected shall be immediately removed by the Contractor. The Contractor must submit the cement, and afford every facility for inspection and testing, at least twelve (12) days before desiring to use it. The Inspector of Cements shall be notified at once upon the receipt of each shipment of cement on the work.

2. **Packages.**—No cement will be inspected or allowed to be used unless delivered in suitable packages, properly branded.

3. **Storage.**—On all main sewers, bridges (unless otherwise ordered), and such branch sewers or other work as the Chief Engineer may designate, shall be provided a suitable house for storing the cement.

4. **Protection.**—Accepted cement, if not used immediately, must be thoroughly protected from the weather, and never placed on the ground without proper blockings.

5. **Failure.**—The failure of a shipment of cement , on any work to meet these requirements may prohibit further use of the same brand on that work.

The acceptance of a cement to be used shall rest with the Chief Engineer, and will be based on the following requirements:

NATURAL CEMENT.

1. **Specific Gravity and Fineness.**—Natural cement shall have a specific gravity of not less than 2.7, and shall leave, by weight, a residue of not more than two (2) per cent on a No. 50 sieve, fifteen (15) per cent on a No. 100 sieve, and thirty-five (35) per cent on a No. 200 sieve; the sieves being of brass-wire cloth having approximately 2400, 10,200, and 35,700 meshes per square inch, the diameter of the wire being .0090, .0045, and .0020 of an inch respectively.

2. **Constancy of Volume.**—Pats of neat cement one-half (½) inch thick with thin edges, immersed in water after "hard" set, shall show no signs of "checking" or disintegration.

3. **Time of Setting.**—It shall develop "initial" set in not less than ten (10) minutes, or "hard" set in less than thirty (30) minutes. This being determined by means of the Vicat needle from pastes of neat cement of normal consistency, the temperature being between 60° and 70° Fahr.

4. **Tensile Strength.**—Briquettes one (1) square inch in cross-section shall develop the following ultimate tensile strengths:

Age.	Strength.
24 hours (in water after "hard" set)............	100 lbs.
· 7 days (1 day in air, 6 days in water)...........	200 lbs.
28 days (1 day in air, 27 days in water).........	300 lbs.
7 days (1 day in air, 6 days in water), 1 part of cement to 2 parts of Standard quartz sand..	125 lbs.
28 days (1 day in air, 27 days in water), 1 part of cement to 2 parts of Standard quartz sand..	200 lbs.

PORTLAND CEMENT.

5. Specific Gravity and Fineness.—Portland cement shall have a specific gravity of not less than 3, and shall leave, by weight, a residue of not more than one (1) per cent on a No. 50 sieve, ten (10) per cent on a No. 100 sieve, and thirty (30) per cent on a No. 200 sieve; the sieves being the same as previously described.

6. Constancy of Volume.—Pats of neat cement one-half (½) inch thick, with thin edges, immersed in water after "hard" set, shall show no signs of "checking" or disintegration.

7. Time of Setting.—It shall require at least thirty (30) minutes to develop "initial" set, under the same conditions as specified for natural cement.

Tensile Strength.—Briquettes of cement one (1) inch square in cross-section shall develop the following ultimate tensile strengths:

Age.	Strength.
24 hours (in water after "hard" set)............	175 lbs
7 days (1 day in air, 6 days in water)...........	500 lbs.
28 days (1 day in air, 27 days in water).........	600 lbs.
7 days (1 day in air, 6 days in water), 1 part of cement to 3 parts of Standard quartz sand..	170 lbs.
28 days (1 day in air, 27 days in water), 1 part of cement to 3 parts of Standard quartz sand..	240 lbs.

9. **Additional Requirements.**—All cements shall meet such additional requirements as to " hot water," " set," and " chemical " tests as the Chief Engineer may determine. The requirements for " set " may be modified where the conditions are such as to make it advisable.

MORTAR.

1. **Sand and Water.**—Sand shall be sharp, silicious, dry-screened, tide-washed bar sand, or approved flint bank sand, free from loam or other extraneous matter. The water must be fresh, and free from dirt. When so directed by the Chief Engineer, salt water may be required, to prevent the mortar from freezing, when absolutely necessary to lay masonry in cold weather.

2. **Composition.**—Portland-cement mortar shall be composed of one part of cement and three parts of sand. Natural-cement mortar shall be composed of one part of cement and two parts of sand.

Mortar for pointing, grouting, bedding coping-stones and bridge-seats, shall be composed of one part Portland cement and two parts sand. A greater portion of cement shall be used when required.

3. **Mixing.** — The ingredients, properly proportioned by measurement, must be thoroughly mixed dry in a tight box of suitable dimensions, and the proper amount of clean water added afterwards. No greater quantity is to be prepared than is required for immediate use, and any that has " set " shall not be retempered or used in any way.

4. **Tensile Strength.**—Mortar taken from the mixing box, and moulded into briquettes one square inch in cross-section, shall develop the following ultimate tensile strengths:

Age.	Strength.
7 days (1 day in air, 6 days in water), 1 part of natural cement to 2 parts of sand...........	50 lbs.
28 days (1 day in air, 27 days in water), 1 part of natural cement to 2 parts of sand...........	125 lbs.
7 days (1 day in air, 6 days in water), 1 part of Portland cement to 3 parts of sand..........	125 lbs.
28 days (1 day in air, 27 days in water), 1 part of Portland cement to 3 parts of sand..........	175 lbs.

PHILADELPHIA, January 9, 1897.

www.ingramcontent.com/pod-product-compliance
Lightning Source LLC
Chambersburg PA
CBHW021516210326
41599CB00012B/1271